Discount Price: 103²⁰

- **Credit Cards preferred**
- **Springer ships internationally**
- **Price excludes tax**
- **Price excludes postage**

This display copy is reserved for:

_____

Please order here with discount.

Geoparks of the World

Development and Management

Spectacular geomorphological landscapes and regions with special geological features or mining sites, are becoming increasingly recognized as critical areas to protect and conserve for the unique geoscientific aspects they represent and as places to enjoy and learn about the science and history of our planet. More and more national and international stakeholders are engaged in projects related to "Geoheritage", "Geoconservation", "Geoparks" and "Geotourism" and are positively influencing the general perception of modern Earth sciences.

Most notably, "Geoparks", have proven to be excellent tools to educate the public about "Earth Sciences". And shown to be areas for recreation and significant sustainable economic development through geotourism.

In order to develop further the understanding of Earth sciences in general and to elucidate the importance of Earth Sciences for society the Geoheritage, Geoparks and Geotourism (G3) Conservation and Management Series has been launched together with its sister GeoGuides series.

'Projects' developed in partnership with UNESCO, World Heritage and Global Geoparks Networks, IUGS and IGU, as well as with the 'Earth Science Matters' Foundation, are welcome.

The series aims to provide a place for in-depth presentations of developmental and management issues related to Geoheritage and Geotourism as well as existing and potential Geoparks. Individually authored monographs as well as edited volumes and conference proceedings are welcome in this series.

This book series is considered to be complementary to the Springer-Journal "Geoheritage".

Kyung Sik Woo · Young Kwan Sohn · Ung San Ahn
Seok Hoon Yoon · Andy Spate

Jeju Island Geopark—A Volcanic Wonder of Korea

Springer

Kyung Sik Woo
Department of Geology
Kangwon National University
Chuncheon, Republic of Korea

Seok Hoon Yoon
Department of Earth and Marine Sciences
Jeju National University
Jeju-si, Republic of Korea

Young Kwan Sohn
Department of Earth and Environmental Science
Gyeongsang National University
Jinju, Republic of Korea

Andy Spate
Optimal Karst Management
Bentley DC, WA
Australia

Ung San Ahn
Jeju Island Geopark Promotion Team
Jeju Special Self-Governing Province
Jeju-si, Republic of Korea

ISBN 978-3-642-20563-7
DOI 10.1007/978-3-642-20564-4

ISBN 978-3-642-20564-4 (eBook)

Library of Congress Control Number: 2012955264

© Springer Verlag Berlin Heidelberg 2013
This work is subject to copyright. All rights are reserved, whether the whole or part of the material is concerned, specifically the rights of translation, reprinting, reuse of illustrations, recitation, broadcasting, reproduction on microfilm or in any other way, and storage in data banks. Duplication of this publication or parts thereof is permitted only under the provisions of the German Copyright Law of September 9, 1965, in its current version, and permission for use must always be obtained from Springer. Violations are liable to prosecution under the German Copyright Law.
The use of general descriptive names, registered names, trademarks, etc. in this publication does not imply, even in the absence of a specific statement, that such names are exempt from the relevant protective laws and regulations and therefore free for general use.

Printed on acid-free paper

Springer is part of Springer Science + Business Media (www.springer.com).

Contents

1	**Introduction**	1
2	**Geographic Setting**	3
	2.1 Climate	3
	2.2 Natural landscapes	5
3	**Habitats**	7
	3.1 Terrestrial ecosystems	7
	3.2 Marine life	7
4	**History**	9
	4.1 Prehistory	9
	4.2 Tamna State	9
	4.3 Goryeo Dynasty	9
	4.4 Joseon Dynasty	9
	4.5 Modern and Present Age	10
5	**Geology of Jeju Island**	13
6	**Geosites**	15
	6.1 Mt. Hallasan Geosite Cluster	15
	6.1.1 Geoheritage	15
	6.1.2 Cultural Heritage	15
	6.1.3 Historical Heritage	17
	6.1.4 Biological Heritage	18
	6.1.5 Legends	19
	6.2 Seongsan Ilchulbong Tuff Cone Geosite	20
	6.2.1 Geoheritage	20
	6.2.2 Historical Heritage	22
	6.2.3 Biological Heritage	23
	6.2.4 Legend of Deunggyeongdol	24
	6.3 Manjang Cave (lava tube cave) Geosite	24
	6.3.1 Geoheritage	24
	6.3.2 Historical Heritage	25
	6.3.3 Biological Heritage: Bijarim (Natural Monument No. 374)	26
	6.3.4 Legend of Gimnyeongsagul (Gimnyeong Snake Cave)	26
	6.4 Seogwipo Formation and Cheonjiyeon Waterfall geosites	27
	6.4.1 Geoheritage	27
	6.4.2 Biological Heritage	29
	6.4.3 Historical Heritage	31
	6.4.4 Legends	31
	6.5 Jisagae Columnar-Jointed Lava Geosite	32

v

	6.5.1	Geoheritage	32
	6.5.2	Cultural Heritage	32
	6.5.3	Historical Heritage	34
	6.5.4	Biological Heritage	34
6.6	Sanbangsan Geosite		35
	6.6.1	Geoheritage	35
	6.6.2	Cultural Heritage	37
	6.6.3	Historical Heritage	39
	6.6.4	Legend	41
6.7	Suweolbong Geosite		42
	6.7.1	Geoheritage	42
	6.7.2	Cultural Heritage	46
	6.7.3	Archeological and Historical Heritages	46
	6.7.4	Legend of Suweolbong	48
6.8	Other Cutural Heritages		48
	6.8.1	Dolhareubang	48
	6.8.2	Doldam (Stone Walls)	48
	6.8.3	Jeju Chilmeoridang Yeongdeunggut	49
	6.8.4	Haenyo	49
	6.8.5	Bangsatap (Stone Towers)	49
	6.8.6	Samseonghyeol	50
	6.8.7	Prehistoric Remains in Samyang-dong	50
	6.8.8	Jejumok Government Office Buildings	50

7 Future Geosites ... 51

7.1	Dangsanbong Tuff Cone	51
7.2	Chagui Island	51
7.3	Biyang Island	51
7.4	Sangumburi Crater	52
7.5	Geomunoreum Scoria Cone	52
7.6	Dusanbong Tuff Cone	54
7.7	Udo Tuff Cone (Someorioreum)	54
7.8	Oedolgae	57
7.9	Songaksan Tuff Ring	58
7.10	Dansan Tuff Ring/Cone	59
7.11	Seopjikoji	60

8 Geotourism ... 61

8.1	Education Facilities		61
	8.1.1	Visitor Centers and Visitor Points	61
8.2	Educational Tourism		65

9 Economy and Development of Sustainable Tourism ... 69

9.1	Economy		70
9.2	Sustainable Development		70
	9.2.1	History of Tourism on Jeju Island	70
	9.2.2	Development of Sustainable Tourism	70
9.3	Socioeconomic Development		71
	9.3.1	Jeju Tourism Organization and Jeju Special Self-Governing Provincial Tourism Association	71
	9.3.2	Geomunoreum Trail Run by Local Residents	71
	9.3.3	Partnership	71

10 Management Plan ... 75

10.1	Purpose of the Management Plan	75

	10.2	Legal Basis for Protection and Management	75
	10.2.1	Legal Basis for Protection	75
	10.2.2	Legal Basis for Management	77
10.3		Management Structure	77
10.4		Potential Pressures on the Geosites	78
	10.4.1	Natural Processes	78
10.5		Current Management Practices and Facilities	79
10.6		Future Action Plans	79
	10.6.1	Developing Visitor Centers	80
	10.6.2	Promotion of Geopark and Geosites	80
	10.6.3	Promoting Education	80
	10.6.4	Developing Partnerships	80
	10.6.5	Involvement of the Community and Non-Government Organizations	81
	10.6.6	Code of Ethics for Visitors and Researchers	81
	10.6.7	Training of Managers and Guides	81
	10.6.8	Promoting Research	81
10.7		Monitoring Indicators and Monitoring Plan	81

References 87

Introduction

1

Jeju Island essentially consists of one major shield volcano, Hallasan (= Mt. Halla, 1,950 m in altitude), with satellite cones building out around its flanks, with some preferential occurrence to the north-east and south-west that explains the elongation of the Island. The islets are remnants of past eruptive cones or lava flows. Hallasan is one of Korea's three sacred mountains and a national park, was established in 1970 to protect its natural and cultural values, to provide for education and enjoyment and to recognize the mountain's spiritual values to the Korean people. As well as the sacred nature and other heritage values of Hallasan, the island has a wide range of natural and cultural features including well-preserved prehistoric sites such as Gosanri dating back to ~12,000~8,000 years before present. There is also evidence of occupation as much as 80,000 years ago. Recognition of the natural and cultural values of Jeju is demonstrated internationally by the inscription of parts of the island as a UNESCO Man and the Biosphere Reserve (in 2002) and as the Jeju Volcanic Island and Lava Tubes World Heritage Area (in 2007). National recognition is based on a number of national parks nature reserves and national monuments as well as a range of provincial protected areas. The nine geosites (including one geocluster) identified and described herein are protected and managed under the national laws of the Republic of Korea and by the Jeju Special Self-Governing Province under many specific laws. First of all the Constitution of the Republic of Korea defines the protection and transmission of traditional and national culture as the responsibilities of the country (The Constitution of the Republic of Korea, Article 9). Therefore the protection of cultural heritage is the fundamental responsibility of the nation and the law that specifies this is the Cultural Heritage Protection Act 2007. All the nine geosites are Korean National Natural Monuments, and the legal basis for management ultimately lies with the national Cultural Heritage Protection Act 2007. This Act is supplemented by regulations and by provincial ordinances and regulations, and by Administrative Directives from the Cultural Heritage Administration. The Jeju Special Self-Governing Province, local businesses, community organizations and academic institutions have established the Jeju Island Geopark and became a member of the Global Network of National Geoparks assisted by UNESCO in 2010.

The Island has a vibrant and developing economy based largely on a tourism industry which already has a strong geotourism emphasis based upon the Jeju Volcanic Island and Lava Tubes World Heritage property supported by many other natural and cultural heritage properties. Jeju Island has a long culture that has long been associated with geological features. As long ago as 1526, a small geomorphic feature was set aside as a protected area with linked ceremonial activities which continue today. This feature, Samseonghyeol; Cave of the Three Clans (See Box below), became associated with legends of the origin of the Jeju people and culture. The ubiquitous stone walls, houses, towers, beacons and, above all the extensive use of Jeju's volcanic rocks for many different carvings of men, animals and mythical beasts are a testament to the Jeju Island long-term cultural involvement with geology. In addition to the Man and Biosphere, World Heritage property, and Global Geopark, Jeju has been designated:

- An Island of Peace (by the Republic of Korea in 2005);
- A Healthy City (by the World Health Organization in 2005); and
- A Free International City (by the Republic of Korea in 2002) with no visa requirements and no taxes.

In September 2009 the Jeju Chilmeoridang Yeongdeunggut festival was inscribed as UNESCO Important Intangible Cultural Heritage No. 71. There are more than 360 registered heritage properties with significant natural and cultural values as well as a large range of natural and developed tourism sites and activities. The challenge for the Jeju Island Geopark, which is accepted by the Jeju Special Self-Governing Province, is to develop and implement strategies that will create an environment that encourages and rewards best practice and sustainable tourism and economic development on the island. These strategies will emphasize the links, geological, social, educational and cultural, between the identified geosites, other natural and cultural heritage sites and the many tourist attractions. This book describes the significant geological features (including 9 geosites) and non-geological heritages. Also, future potential geosites are listed and briefly introduced. Finally, geotourism, sustainable development and management issues are included.

K. S. Woo et al., *Jeju Island Geopark—A Volcanic Wonder of Korea,* Geoparks of the World,
DOI 10.1007/978-3-642-20564-4_1, © Springer Verlag Berlin Heidelberg 2013

Fig. 1.1 Samseonghyeol Cave of the Three Clans

Samseonghyeol—Cave of the Three Clans (Historic Sites No. 134)

This site, with its distinct depression and traces of three holes, demonstrates the ongoing link between geology and culture so vividly evident on Jeju. Samseonghyeol, located in Ido-dong, Jeju City, is the mythical birthplace of the Jeju people and their culture. It is the sacred site of the myth of the three clans which tells a story of three demi-gods, who came from within the earth and founded the Tamna Kingdom. There are still three holes remaining, which are believed to be the birthplace of these three demi-gods. The people of Jeju have passed the myth of the three clans down over the centuries by word of mouth. It is also found from a couple of ancient documents, such as Goryeosa, Yeongjuji, etc. The myth of the three clans is as follows: in the beginning man did not exist in Jeju. All of a sudden, one day, three demi-gods named Eulla rose from Samseonghyeol and became the originators of three clans, Goh, Yang and Boo families. They lived in harmony, sharing hunting within their carnivorous culture. One day a box was discovered drifting down from the East Sea. On opening the box they discovered three princesses from Byeokrang (a fictional state) in the East Sea with a calf, a colt, and the seeds of five individual grains. Each of the demi-gods married one of the princesses; according to each one's age. They each shot an arrow, settled wherever it landed, and began planting crops and raising animals. They soon prospered in their new agricultural economy and the island developed into the Tamna Kingdom (Fig. 1.1).

Fig. 1.2 Dolhareubang, the most traditional stone statue of Jeju Island

Geographic Setting

2

Jeju is an island, volcanic in origin, situated on the continental shelf 90 km south of the Korean Peninsula. Specifically, Jeju Island is located between 33°11′27″ and 33°33′50″ north in latitude and between 126°08′43″ and 126°58′20″ east in longitude. The island is a slightly flattened ellipse, ~70 km in length from south south-west to north north-east, and varying from 30~35 km in width. In addition there are a number of rocky islets just offshore. The east side of Jeju faces Tsushima and Nagasaki in Japan across the South Sea of the Korean Peninsula and the East China Sea. The west side of the island faces Shanghai in China across the East China Sea. Jeju is 450 km from Seoul, 270 km from Busan, 330 km from Fukuoka in Japan, and 500 km from Shanghai in China.

Jeju Island is a province of the Republic of Korea; it has singular status as the Jeju Special Self-Governing Province. It is divided into two Administration District and further divided into 43 smaller units variously termed 'eup', 'myeon' or 'dong'. There are two major cities and 556 natural villages large and small. It has an area of 1,814 km² (Figs. 2.1 and 2.2).

2.1 Climate

Located in eastern region of the continent, and in the mid-latitudes of the Northern Hemisphere, Jeju has strong characteristics of east-coast climate, experiencing clear changes of season. Because of its geographic settings as a lone island off in a distant sea, Jeju experiences relatively short winters and longer summers. While it is affected, as other areas in Korea, by the northwest wintertime monsoon, Jeju is also affected by the southwest and southeast monsoons in summertime. Hallasan, being in the center of the island, causes climatic characteristics of each region. Climatic differences of the surrounding seawaters are responsible for significant influences on various geographical features and also on the lives of the regions inhabitants. According to Köppen's classification of climate, the climate of Jeju is classified as a subtropical humid climatic regime (Köppen Cfa) so that most areas, except mountainous regions, maintain a mild winter with relatively consistent precipitation.

The average annual temperature of Jeju is 15.5 °C; the average temperature in August, the warmest month, is 26.5 °C; the average temperature in January, the coldest month, is 5.6 °C. The average annual temperature of the city of Gosan, located in the west of Jeju City, is 15.5 °C; the average temperature in the warmest month of August is 26.1 °C; the average temperature of the coldest month of January is 6.3 °C. Gosan has higher temperatures than Jeju City during winter because of strong winds in Gosan that prevent cooling, while lesser wind activity in Jeju results in more active cooling. The temperature regime is influenced by the nearby oceanic currents.

The precipitation of Jeju is influenced mainly through cyclone activity: precipitation of seasonal rain takes place mostly in summer months and precipitation caused by typhoons takes place in both summer and autumn. There is also a very limited amount of precipitation caused by Siberian anticyclone activity in wintertime. Precipitation is largely caused by southwest and southeast air currents. In winter time precipitation is often caused by the northwest air current. Jeju has the highest annual precipitation of Korea ranging from 1,000 to 1,800 mm. The average annual precipitation at is 1,457 mm; Jeju City and 1,851 mm at Seogwipo. Seasonal variations in the precipitation of Jeju Island are wide: 47 % of the rain falls in summer, while 12 % falls in winter.

Jeju is famous for its wind. Wind velocities are very high and the daily frequency is consistent. The average annual wind velocity of Jeju City is 13.6 kph and Gosan's is 24.8 kph; this compares with Seogwipo's and Seongsanpo's 11.2 kph. The northwest slope of Hallasan maintains constant and significantly stronger wind velocities. Gosan, in particular, records average wind velocities at more than 50 kph for a period of at least 80 days compared to Jeju City's 14.5 days, Seogwipo's 2.8, and Seongsanpo's 0.9 days. The most important wind in Jeju Island is the northwest monsoon in wintertime. The average annual wind velocity in Gosan reaches 33.5 kph, whereas Jeju reaches 16.9 kph. In terms of the wind direction, Jeju maintains a high frequency of northwest

K. S. Woo et al., *Jeju Island Geopark—A Volcanic Wonder of Korea*, Geoparks of the World, DOI 10.1007/978-3-642-20564-4_2, © Springer Verlag Berlin Heidelberg 2013

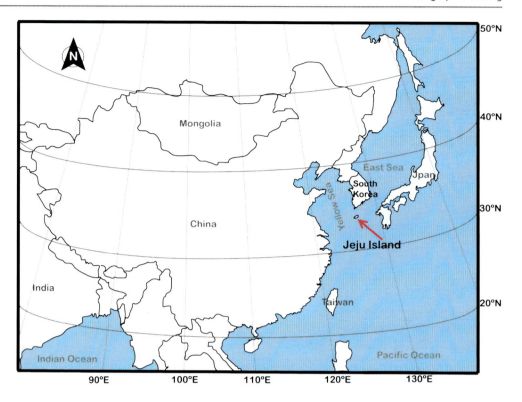

Fig. 2.1 Location of Jeju Island

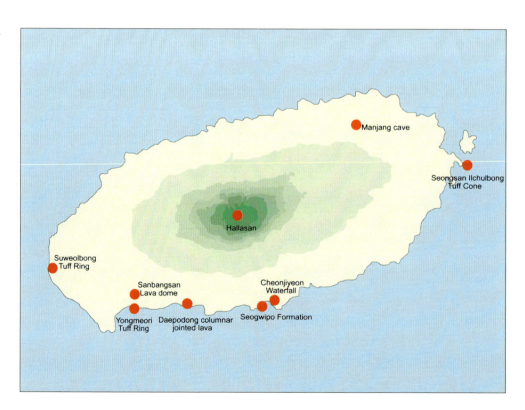

Fig. 2.2 Location of nine geosites in the Jeju Island Geopark

2.2 Natural landscapes

Fig. 2.3 Aerial view of the Seongsan Ilchulbong Tuff Cone and Hallasan in the back. Both are the World Heritage sites

Fig. 2.4 Scenic view of a trachyte dome at the peak with azaleas blooming in spring

winds in winter and east winds in summer. The strong winds in Jeju have had a significant influence on the natural environment and on the lives of its inhabitants.

2.2 Natural landscapes

Jeju Island was formed from volcanic activity occurring during the Quaternary period. As a result, the drainage system, mountain system and coastal topography of Jeju show specific characteristics related to how and when the volcanic activities occurred. The plan of Jeju is an oval with a major axis of N70°E, which parallel with the south coastline of the Korean Peninsula, and is accord with NE-SW tectonic lines on the Korean Peninsula. The island as a whole formed from a shield volcano with Hallasan as a central, and main, erupted center. The coastline, approximately 253 km long, is generally unvaried and mostly made up of rocky shores exposed volcanic rock, and occasionally small-sized pocket beaches with a limited number of sand dunes in some areas.

Fig. 2.5 Scenic view of trachybasalt columns (Five Hundred Diciples of Buddha) in Hallasan

The overall geomorphologic features of Jeju Island are, in large part, divided into three categories as follows:
- Lava plateaus developed in lower parts of coastal areas;
- The shield volcano of Hallasan in the center of the island; and
- Volcanic cones and craters (oreums in the Jeju dialect) surrounding Hallasan.

At the summit of Hallasan, there is a crater lake named Baeknokdam with maximum and minimum width of 585 and 375 m, respectively. About 360 volcanic craters and cones lie scattered about the major axis of the island so that they dominate the overall topography and scenery of Jeju. Volcanic activity along the coastal regions had occurred in a watery environment, causing the formation of tuff rings and cones, and later on the constant seawater erosion continued to create unique and ever-evolving coastal landscapes. Another geomorphologic feature of Jeju Island is the large-scale lava tubes which have developed underground. Lava of low viscosity and high fluidity flowed repeatedly toward the ocean from the volcanic centers around Hallasan so that world-scale lava tubes were created beneath the surface.

Jeju's drainage system is made up of streams that radiate outwards from the central high-point of Hallasan. While drainage systems created from wide lava plateaus formed on the gentle east-west slope of Hallasan are less evolved, most watercourses were developed on the north-south slopes and run either southbound or northbound. Owing to nature of the geological features there are no permanently running streams on Jeju. Water from upper streams runs into the underground through permeable layers developed on the edges of lava flows or along columnar joints in the streambeds. Although the ground water rises to the surface near the seashores, sometimes forming waterfalls, most of water courses in Jeju are dry stream beds for much of the year (Figs. 2.3, 2.4 and 2.5).

Habitats

3

3.1   Terrestrial ecosystems

Jeju Island, a volcanic island, is in an oceanic climatic zone and has a relatively mild weather with 15.5 °C as an average annual temperature. However, the climate of Hallasan varies according to altitude so that the geographical distribution of subtropical and arctic plants and animals is different. The land ecosystems of Jeju, reaching from the seaside tidal zone to Baeknokdam, on Hallasan, can be divided into six domains—a coastal wetland zone, an evergreen broadleaved forest zone, a grassland zone, a deciduous forest zone, a coniferous forest zone and an alpine shrub zone.

The coastal wetland zone is mainly distributed on a small scale in the region close to the seaside tidal zone where water gushes out from seaside springs. The dominant species are *Pinus thunbergii* and halophilic plants such as *Crinum asiaticum* var. *japonica*, Hamabo mallow, *Paliurus ramosissimus*, *Ixeris repens* and various reeds. Migratory birds include Black-faced Spoonbills, Whooper Swans, Storks, Mandarin Ducks, and Herring Gulls.

The evergreen broadleaved forest zone is characterized by many sites such as the Cheonjiyeon and Cheonjeyeon valleys with *Castanopsis cuspidate* var. *sieboldii*, *Elaeocarpus sylvestris* var. *ellipticus*, *Cymbidium kanran*, *Cymbidium goeringii*, and *Ardisia japonica* found therein. There are many permanent resident birds such as Brown-eared Bulbuls, Japanese White-eyes, Great Tits and Bush Warblers.

The grassland zone is largely successional arising from past disturbance. A small wind rose, *Smilax china*, and *Pinus thunbergii* are prominent. Small marshes form sporadically with marsh-land plants such as *Isoetes japonica*, *Brasenia schreberi*, *Marsilea quadrifolia* and animals such as Black-spotted Pond Frogs and Tiger Keelback Snakes.

The deciduous forest zone includes areas such as Seongpanak, Eorimok, Yeongsil, and Tamla valley on Hallasan. Dominant tree species include *Prunus yedoensis*, *Carpinus laxiflora*, *Sasaquelpaertensis* sp., *Ligularia fischeri*, and *Polygonatum odoratum* var. *pluriflorumas* as well as animals such as Roe Deer, Badgers, White-backed Woodpeckers, Great Tits and Jays.

In the coniferous forest zone, dominant tree species include a large number of *Abies koreana*, *Pinus densiflora*, *Taxus cuspidata* and *Juniperus chinensis* var. *sargentii*. In addition, shrubs such as Rhododendron mucronulatum var. ciliatum and Crowberry are widely distributed in this buffer zone to Baeknokdam. Roe Deer, Jungle Crows, Coal Tits, Common Buzzards and Peregrine Falcons can be found here. In the alpine shrub zone, there are many classes of shrubs of small size on account of a low temperature and strong winds. These include *Rhododendron mucronulatum* var. *ciliatum*, *Diapensia japonica* var. *obovata* and *Betula schmidtii*. Jeju Salamanders and the Korean Fire-bellied Toads inhabit the Baeknokdam lake.

3.2   Marine life

Jeju Island is geographically located within a temperate and subtropical zone, and is naturally surrounded by the ocean. The total length of the coastal line is 253 km with a broad continental shelf located 100 m below the sea level. Comparatively high winter water temperatures allow a variety of fish to use this area for spawning and overwintering. Accordingly, many abundant fishing grounds are formed by the warm currents from the East and West Seas of Korea. Three hundred and fifty species of sedentary and migatory fish inhabit the waters. Rare fish such as Whale-sharks, Basking-sharks, Devil-rays, Japanese diamond-skates, Manchurian sturgeons and Oarfish also occur.

With the exception of some sandy coasts, most of the coastal area is comprised of rock beds which provide a good habitat for marine algae and mollusks. The flora of marine algae area is varied and has been classified as the Jeju Zone. The marine algae flora is subdivided as follows: 2 % belonging to the northern type, 10 % southern type, 74 % temperate type in addition to 14 % indicating characteristics of the island. Also of special interest are some 150 species of shellfish such as abalone and turban shells which find algae to be their main food source. Also notable within this community are 20–25 species of shellfish found along the sandy beaches

Fig. 3.1 *Diapensia lapponica* var. *obovata*, an arctic shrub at the summit of Mt. Hallasan

Fig. 3.3 Cones of the Korean fir (*Abies koreana*)

Fig. 3.2 Korean fir forest near the summit of Mt. Hallasan

Fig. 3.4 *Empertrum nigrum* var. *japonicum*, These temperate deciduous hardwoon forests and evergreen coniferous forests are the major vegetation types of the Hallasan Natural Reserve

Fig. 3.5 Roe deer near the trail on Mt. Hallasan

of Kwakji-ri and the Seongsan area. More than 150 species of crustacean and other sea creatures can be found on Jeju including lobster shrimp, crab and abalone as well as starfish. Reptiles, cephalopods, sea urchins and sea cucumber are also abundant as are corals [Figs. 3.1, 3.2, 3.3, 3.4 and 3.5].

History

4

4.1 Prehistory

The history of Jeju began during the Paleolithic Age, 70,000–80,000 years ago. Jeju people from the prehistoric age mostly lived in caves. In Billemotgul (cave) there are Paleolithic artifacts, including chipped stone tools and bones of reindeer and bear which are today found to inhabit only Siberia or Alaska. Prehistoric remains in Gosan-ri, Hangyeong-myeon are the oldest remains from the Neolithic Age in Korea, dated 8,000–12,000 years ago. Hunting tools such as arrowheads, spearheads and various earthenware types excavated from the site show the methods and practices of those surviving by hunting and food-gathering within a group.

4.2 Tamna State

The ancient name of Jeju, Tamna, first appeared in a written document during the 6th century. This indicates that Jeju practiced a settled ruling system and was able to unite the society as a maritime island nation unique from neighboring areas. Such a political and social system is said to have initially appeared in Jeju in the time of the third to second centuries BC. The tale of the three family names also originated from that period, telling the story of three demi-gods who rose from Moheunghyeol, or Samseonghyeol, married three princesses from the fictional state of Byeokrang and began planting crops and raising animals. Archaeological research discovered Samyang-dong remains in Jeju revealing large and small dugout huts, a storehouse, a stone embankment, and a drainage system with more than 240 dwelling structures. They are the largest ancient village remnants in South Korea and are from the period between 200 BC and 200 AD. It shows that there was a large-scale village already prospering during the formative period of the Tamna State. From the Three Kingdoms period, the Tamna State (officially named as Takra, or Tamna) started trading with Baekje, Goguryeo and Silla on the Korean Peninsula. After Baekje was destroyed by Silla-Tang allied forces in 660 AD, the Tamna State began diplomatic relationships with Japan and the Tang Dynasty in China.

4.3 Goryeo Dynasty

Once the Goryeo Dynasty was established on the Korean Peninsula, the Tamna Crown Prince would be sent to the Goryeo Court. Around the twelfth century the Tamna State was annexed to the Goryeo central government, officially named Jeju, and was organized into administrative units termed gun, hyeon, mok and so on. When Mongolia attacked Goryeo in the thirteenth century, Jeju became the last base of the Sambyeolcho Army which earned a reputation as never surrendering to the Mongols. After defeating the Sambyeolcho Army, Mongolia set up Tamna Chonggwanbu and directly ruled Jeju for 20 years. Intense horse-breeding began in Jeju at that time with the introduction of 160 head of horses from Mongolia. Although Jeju was returned to Goryeo rule in 1294, the settled Mongolian community exercised their influence until 1374 when General Choi Yeong suppressed an uprising by Mokho, the Mongolian shepherds.

4.4 Joseon Dynasty

In the early Joseon Dynasty period, following the Goryeo Dynasty, Jeju established the one-mok and two-hyeon system. This included Jeju-mok, Daejeong-hyeon and Jeongeu-hyeon and maintained a fast-growing population increasing from ten thousand in the Goryeo Dynasty, to over sixty thousand by the time government offices and its castle were built. During the fifteenth to seventeenth centuries, however, many years of famine caused a population decline prompting the Joseon Court to enact a law forbidding the Jeju people leaving the island. In addition, due to its unique location as a lone island in a distant sea, Jeju was the locale where more than two hundred people lived in exile during the 500-year long Joseon Dynasty. Social positions of exiles varied including a dethroned king, Gwanghaegun, a royal family, a

Fig. 4.1 Various stone statues of old men of the Hallim Park, a potential geosite.

Fig. 4.3 A traditional wedding at the Jejumok government offices, the hub of politics, administraton and culture of Jeju Island during the Joseon Dynasty

Fig. 4.2 Daejeong Hyanggyo, a local educational institution in the Joseon Dynasty

Fig. 4.4 Jeju Harbour in 1890

politician, a scholar, a monk, a eunuch and a thief. While some of them were given the death penalty by poison or were transferred to other regions, some exiles settled down in Jeju after being pardoned with each becoming the originator of their clans (Figs. 4.1, 4.2, 4.3 and 4.4).

4.5 Modern and Present Age

Under Japanese rule (1910–1945), the "gun" regional system replaced the "do" regional system of Jeju in 1915. The circular road along the coasts was opened in 1917. Anti-Japanese movements lasted from 1918 until 1932. In particular, the Haenyo (women divers), on the eastern area of the island, raised the largest anti-Japanese movement by woman in Korea, with 17,000 participants. When the United States Army intensified its attack, the Japanese Army built up more than 90 positions on major oreums and coastal areas through the mobilization of Jeju laborers. After the liberation of Korea from Japanese rule, Jeju had its own local government independent from Jeollanam Province. In 1948 the April 3rd Massacre occurred in which tens of thousands of citizens were killed by an ideological conflict between the leftwing and rightwing political factions. In 2000, Jeju was designated as an 'Island of World Peace' owing to the wishes of its people that peace and humanity will overcome the tragedy of the April 3rd Massacre. In 2006 the Jeju Special Self-Governing Province, including two administrative cities, seven eups, five myeons, and thirty-one dongs, was launched in order to develop Jeju into a competitive free international city (Figs. 4.4 and 4.5).

4.5 Modern and Present Age

Fig. 4.5 Haenyeo, traditional women divers of Jeju Island, and the Seongsan Ilchulbong Tuff Cone behind

Jeju has an ancient tradition involving stone.
Jeju – island of Samda and Sammu
From ancient times Jeju-do has been called Samda-do. *Samda* means three plentys. That is, *plenty of stones, of wind, and of women*. *Plenty of stones* originated from the volcanic activities of Hallasan back in olden times. Jeju people have improved their lives by going through a lengthy process of reclaiming wasteland comprised of stones and constructing a port with barriers at the seashore. *Plenty of wind* also represents unusually harsh lives in Jeju. People had to fight against the rough sea with typhoons constantly passing through. Along with *plenty of stones, plenty of wind* had a broad effect on people's lives: Jeju people built their houses with a stone fence placed low to the ground, made thatched roofs and fastened ropes across over the top. Another example of note is the farmlands with field fences made of stones. *Plenty of women* is based on the fact that men died at sea while fishing. More importantly, however, in order to compensate, Jeju women have a long history of assuming the duties of men in order to survive the harshness of life there. Although *plenty of women* demonstrates the actual ratio of woman to men in Jeju, it is based more as a metaphor reflecting the character of the hardworking Jeju women. Haenyeo, the Jeju woman diver, is the symbol of the island; all women work in the sea, fighting against the rough waves.

Sammu stands for three *absences*. That is, *absence of thieves, of beggars, and of gates*. Jeju people have had to live modestly, applying a philosophy of resilience and persistence while being aware of others and offering support in order to survive the harshness of nature. Within the rough, natural lifestyle they also kept aware and offered mutual assistance in order to avoid the existence of the deprived and of thieves. From this, Jeju is absent of gated houses. Moreover, Jeju people rarely participated in questionable, shameless or poor behavior. It must be noted how this lifestyle grew from not only trying hard to maintain dignity and the respect of their proud ancestors but it was also born out of the fact that Jeju is a very small island and everyone knew on another. The Jeju people's independent life of self-reliance and high esteem did not demand gates to bar entry. An owner of a house only had to place wooden bar at the entrance: this wooden bar is the 'Jeongnang' of Jeju.

Geology of Jeju Island

Jeju Island is a volcanic island situated off the southern coast of the Korean Peninsula. The island was produced by volcanic activity which occurred from about 2 million years ago until historic times. The island is 73 km long in the east-west direction and 31 km long in the north-south direction, having an area of 1,847 km^2. The island has the typical morphology of a shield volcano, characterized by an overall gentle topography and an elliptical shape in plan elongated in the ENE direction (Fig. 5.1).

Basaltic to trachytic lavas occur extensively on the island together with diverse volcanic landforms, including Mt. Hallasan that rises 1,950 m above sea level at the center of the island and about 360 volcanic cones that are scattered throughout the island (Sohn and Park 2007) (Fig. 5.2). In the subsurface, however, numerous hydromagmatic volcanoes (tuff rings and tuff cones) produced by explosive hydrovolcanic activity occur extensively together with intervening volcaniclastic sedimentary deposits (Sohn and Park 2004; Sohn et al. 2008). This is because the volcanic activity of the island commenced in the continental shelf in the southeastern Yellow Sea where abundant water for hydrovolcanic explosion was available. Composed of an early-stage product of hydrovolcanism and a late-stage product of lava effusion, Jeju Island can be defined as a "shelfal shield volcano", distinguished both from nonmarine shield volcanoes and from oceanic volcanic islands that were built on the deep ocean floors.

Thousands of groundwater bores have been drilled all over the island since the 1960s, greatly improving our understanding of the surface and subsurface geology of Jeju Island (Fig. 5.3). The basement is composed of granite and silicic volcanic rocks of Jurassic to Cretaceous age (Kim et al. 2002). The overlying U Formation is 70–250 m thick and composed of well-sorted, quartzose sand and mud (Koh 1997) that were deposited on the continental shelf before the onset of volcanism at Jeju area (Sohn and Park 2004). The U Formation is overlain by about 100 m of basaltic volcaniclastic and fossiliferous deposits named the Seoguipo Formation. Recent studies (Sohn and Park 2004; Sohn et al. 2008) reveal that the formation is composed of numerous superposed phreatomagmatic volcanoes intercalated with marine or nonmarine, volcaniclastic or non-volcaniclastic deposits with intervening erosion surfaces and palaeosol layers. The widespread and continual hydrovolcanic activity together with volcaniclastic sedimentation, as represented by the Seoguipo Formation, is inferred to have persisted for more than a million years (from ca. 1.8 Ma to 0.8–0.4 Ma) under the influence of fluctuating Quaternary sea levels.

Thereafter, proto-Jeju Island (Fig. 5.4) has grown up above the fluctuating Quaternary sea levels and lava effusion became dominant, resulting in the plateau- and shield-forming lavas together with numerous volcanic cones (Fig. 5.2). K-Ar ages of these lavas range generally between 0.8 and 0.03 Ma (Tamanyu 1990; Lee et al. 1994), suggesting that the construction of Jeju Island was almost complete before the Holocene. After the last glacial maximum 18,000 years ago and during the middle Holocene when the coastal regions of Jeju Island became suitable for hydroexplosions, explosive hydrovolcanic eruptions occurred at several places along the present shoreline. These Late Pleistocene to Holocene hydrovolcanic eruptions resulted in several tuff rings and tuff cones with fresh morphology (Sohn and Chough 1989; Chough and Sohn 1990, 1992, 1993). There are also historic records of minor eruptions afterwards about one thousand years ago, although it is uncertain where these eruptions occurred.

Fig. 5.1 Digital elevation model of Jeju Island, showing the overall shield morphology of the island with a central peak (Mt. Hallasan) and numerous volcanic cones

Fig. 5.2 Geological map of Jeju Island covered by numerous lava flows and volcanic cones (Park et al. 2000)

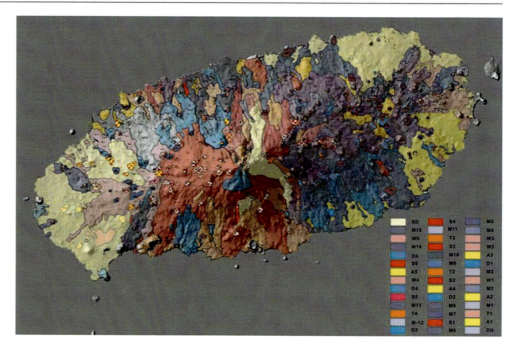

Fig. 5.3 Illustration of the subsurface stratigraphy and lithology of Jeju Island

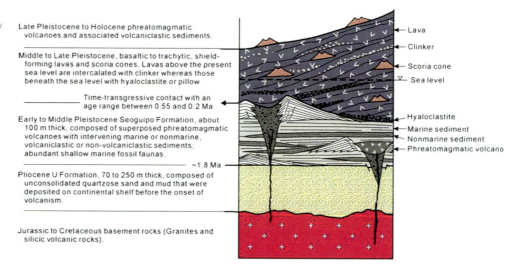

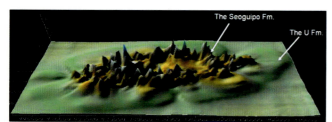

Fig. 5.4 Computer-generated image of the *upper* surface of the Seoguipo Formation, showing the probable appearance of ancient Jeju Island before the effusion of the shield-forming lavas. This image was compiled from about 1,400 borehole data (Sohn and Park 2007)

Geosites

6

6.1 Mt. Hallasan Geosite Cluster

6.1.1 Geoheritage

Mt. Hallasan is the central peak of the gently sloping shield volcano of Jeju Island. It is the highest mountain in South Korea, reaching 1,950 m above sea level. Mt. Hallasan is the symbol of Jeju Island and a representative product of the Quaternary volcanism in the Korean Peninsula and adjacent seas. Mt. Hallasan boasts peculiar volcanic landscape, produced by the crater lake *Baeknokdam* at the summit (Fig. 6.1), the precipitous rocky cliffs of the *Yeongsilgiam* (Fig. 6.2), and about forty volcanic cones. Mt. Hallasan was designated as a natural monument (no. 182) in 1966 and a national park in 1970 because the mountain preserves the pristine morphology of a shield volcano unaffected by significant weathering or erosion. The mountain has been protected from human activity since then and is renowned for its unique ecology and biodiversity in addition to volcanic geology and geomorphology. The mountain was designated as a UNESCO Biosphere Reserve in 2002 and as a UNESCO World Natural Heritage in 2007.

Mt. Hallasan is composed of numerous basaltic to trachytic lavas and a number of volcanic cones (Fig. 6.3). It has a small (108 m deep and 550 m wide) crater at its center instead of a caldera, which is named *Baeknokdam* (Fig. 6.1). Mt. Hallasan is interpreted to have formed since the Middle Pleistocene, after about 780,000 years ago, when the dominantly hydrovolcanic eruption in the early stage of Jeju volcanism was replaced by lava effusion. The volcanic rocks near the *Baeknokdam* crater at the summit were erupted tens of thousands years ago. Because of its young age, the summit area of Mt. Hallasan preserves fresh volcanic landforms and rock formations.

The summit of Mt. Hallasan provides different sceneries when viewed from different directions because the summit area was made from two different lavas that have contrastingly different properties. That is, the western half of that area was made from highly viscous trachyte lava, forming a dome-like topography (Fig. 6.4), whereas the eastern half was made from highly fluid trachybasalt lava, resulting in a gently sloping topography. The southern and northern slopes of the summit are bounded by precipitous rock cliffs because of collapse of the trachytic lava dome (Fig. 6.5). In contrast to the summit area, the flanks of Mt. Hallasan were carved by several deep valleys and gorges. The topographic features formed by erosion and collapse of the volcano are especially well developed in the *Yeongsilgiam* (Fig. 6.2).

6.1.2 Cultural Heritage

6.1.2.1 Grazing Fields of Jeju Horses

The mild climate and vast grasslands of mid-mountain area have made Jeju a perfect place for horse grazing. Jeju Horses (Jorangmal) are rather mid-sized, with the average height of females at 117 cm and males at 115 cm, yet they are mild in temperament and strong enough to resist disease. The majority are brown in color with others being reddish brown, gray and black. Jeju Horses are distinctly smaller than average and have a unique long body shape with a lower front and higher back section.

Jeju is considered to have a long history of raising horses based on the myth of the Three Family Names which relays the story of how three princesses brought the seeds of five grains, a calf and a foal to Jeju. It is recorded that Jeju horses were presented to the king in 1073 during the Goryeo Dynasty. Intense horse-breeding began with the inflow of 160 head of horses from Mongolia in the late 13th century. During the Joseon Dynasty, The mid-slope areas of Halla Mt. between 200 and 600 m above sea level were divided into ten districts. They equipped them with ten grazing fields and started supplying official horses making Jeju famous as a horse-breeding base for many centuries.

Jeju Horses have influenced every aspect of the private sector: the agriculture of Jeju utilized horses mostly for plowing. Jeju Horses were also used for pulling carts and operating millstones. Mabullimje (religious service for the propagation of horses) was performed in the grassland. Jeju has

K. S. Woo et al., *Jeju Island Geopark—A Volcanic Wonder of Korea*, Geoparks of the World,
DOI 10.1007/978-3-642-20564-4_6, © Springer Verlag Berlin Heidelberg 2013

Fig. 6.1 The *Baeknokdam* crater at the summit of Mt. Hallasan

Fig. 6.2 A view of *Yeongsilgiam*, which is composed of columnar-jointed trachyte lavas

the most taboo words and proverbs related to horses and the most advanced popular remedies in Korea. Although it once drastically diminished the number of horses due to limited demand and economic efficiency, today Jeju has more than 20,000 head of horse and is fully equipped for horse racing and horse riding.

In Gyeonwalak grazing field (Yonggang-dong, Jeju), Jeju horses bred by the Jeju Provincial Government graze between May and November every year and visitors are welcome to view the site. Additionally, visitors can review the reference materials of Jeju Horse culture at the Jeju Folklore and Natural History Museum and can enjoy the Jeju Jorangmal race at the Jeju Horse Race Park over the weekends (Fig. 6.6).

6.1.2.2 Jonjaam Site (Jeju Province Monument No. 43)

Located on the slopes of Bolae Oreum, north-east of Yeongsil, this is the site where Jonjaam once stood; this temple was built presumably during the late Goryeo Dynasty or in the early Joseon Dynasty. Excavation research in 1993 and 1994 found that there had been temple buildings and an annex on 4-stair stone grounds. Among the remains unearthed were pieces of celadon porcelain and patterned roof tiles from the late Goryeo Dynasty or the early Joseon Dynasty. A large quantity of white porcelain and non-patterned roof tiles belonging to the early and the middle of Joseon Dynasty were excavated and later restored. Evidence from these excavations suggests Jonjaam was erected during the late

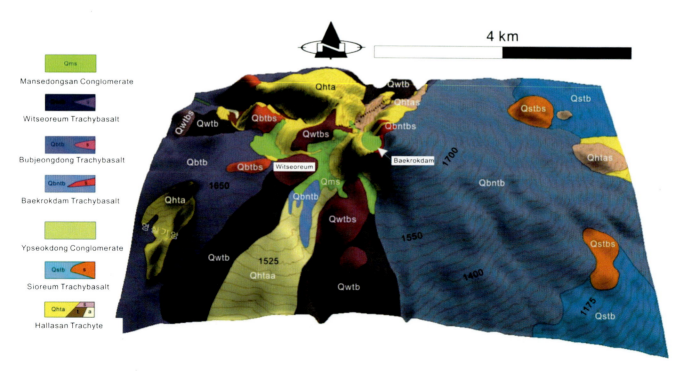

Fig. 6.3 Geological map of Mt. Hallasan, which is composed of five lava units and associated volcanic cones as well as recent volcaniclastic deposits. Topographic contours are in meters

6.1 Mt. Hallasan Geosite Cluster

Fig. 6.4 A lava dome composed of the Hallasan Trachyte occupies the western part of the Baeknokdam crater

Fig. 6.5 The southern part of the Baeknokdam crater forms a precipitous cliff probably because of collapse of the lava dome, which is severely jointed and fractured

Fig. 6.6 Gyeonwalak grazing field of Jeju horses managed by Jeju Provincial Government

Goryeo Dynasty or the early Joseon Dynasty, but conflicting documentation from the Joseon Dynasty, *Jonjaamgi* by Kim Jeongui, states that it was built even earlier when three family names, Goh, Yang and Boo had first arrived. An additional document revealed the existence of a unit for Buddhist discipline teachings near the site. The entire temple had been completely destroyed during the mid-17th century. The present temple buildings were restored in 2002 (Fig. 6.7).

6.1.2.3 Sejonsari Pagoda (Jeju Province Tangible Cultural Property No. 17)

Sari or Sarira is the term given to Buddhist relics; bead-shaped objects found among the cremated ashes of Buddhist monks used during Buddhist faith practices. Sari pagoda is a grave structure which enshrines the Sari relics. The 1993 excavation research at the Jonjaam Site found an oval-shaped Sejonsari Pagoda which was built of Jeju basalt. Sejonsari Pagoda has a round-shaped Joongdaeseong (middle stone) on an octangular stereobate, and on those, a bell-shaped Tapsinseog (main body stone) and Okgyeoseog (roof stone) as a roof, and on the top, a lotus blossom-shaped ornament decorating the body (Fig. 6.7). As the overall shape resembles a bell, Sejonsari Pagoda is categorized as a stone-bell style, which was popular from the late Goryeo Dynasty through the early Joseon Dynasty.

6.1.2.4 Seated Wooden Gwaneumbosal Statue in Gwaneumsa Temple (Jeju Province Monument No. 16)

Gwaneumbosal is a Buddhist saint who saves mankind by granting mercy. The seated wooden Gwaneumbosal statue in the main building of Gwaneumsa in Jeju is one of the most exceptional works, having an elegant full-length body, detailed facial expressions, three-dimensional wrinkles in the garment and pays homage to the accurate and specific characteristics of the Joseon Dynasty Buddhist statue of the late 17th century. Beyond the characteristics of Bosal, a Buddhist saint, it wears an upper garment, a surplice and a splendidly decorated coronet like the Yeorae Buddha. The coronet has a seated Amitabha Yeorae statue carved in the center, as an indication of Gwaneumbosal. The seated statue is 75 cm high and 47 cm wide. Records show it was created in 1698 and placed in Daeheungsa, Jeollanam-do. The statue was moved to Gwaneumsa in Jeju in 1925 (Fig. 6.8).

6.1.3 Historical Heritage: Eoseungsaengak Japanese Military Caves in World War II (Registered Cultural Heritage No. 307)

Eoseungsaengak on the northern slope of Mt. Halla is one of the largest scoria cones in Jeju Island. This was the location for the headquarters of the 58th Japanese Army during

Fig. 6.7 Photos of Jonjaam site during the excavation research in 1993 (**a**) and after the restoration of temple buildings and Sejonsari Pagoda in 2002 (**b**)

Fig. 6.8 Seated wooden Gwaneumbosal statue (**a**) in the main building of the Gwaneumsa Temple (**b**)

the Pacific War (World War II). Formed in April of 1945 the 58th Japanese Army was assigned the mission of preventing the U.S. Armed Forces from attacking the northern city of Kyushu on the Japanese mainland. Japanese Army eventually established a base on Jeju due to its strategic value; it allowed a vantage point of not only the city of Eoseungsaengak but also the entire western and the northeastern areas of Jeju. Eoseungsaengak was a significant military asset with command headquarters for all defensive positions based on the shores and mid-mountain areas. Two Japanese Army bunkers and three cave dug-out positions still remain in the south-western section of Eoseungsaengak. The total length of the cave dug-out positions is approximately 460 m. The 120 troops of the Japanese 96th Anti-aircraft Division, under the command of the 58th Army, remained headquartered in Eoseungsaengak until the conclusion of the Pacific War (Fig. 6.9).

Fig. 6.9 Japanese military remains (pillbox) of WWII at the summit of Eoseungsaengak scoria cone

6.1.4 Biological Heritage

6.1.4.1 Natural habitats of Cherry trees (*Prunus yedoensis*) at Bongae-dong (Natural Monument No. 159)

Prunus yedoensis is a deciduous broad-leaved tree, which grows in rich fertile soil to a height of more than 15-meters and does not grow as well in the cold weather. Its dense, non-twisting timber is good for furniture, tools and interior finishing. Its bark is used for craft products and its fruit is edible. In 1905 French priest Emile Taquet sampled the tree and sent it to Dr. Koehne at the Berlin University of Germany. From this it was discovered that Jeju is the only place throughout the globe where *P. yedoensis* grows. It grows mostly in the Hallasan terrain, 500–900 m above sea level. The natural habitats in Bongae-dong, Jeju and in Shinryeori, Seogwipo, have been preserved as a natural monument.

6.1 Mt. Hallasan Geosite Cluster

Fig. 6.10 Natural habitat of cherry trees (*Prunus yedoensis*) at Bongae-dong. Photos taken in the spring (**a**) and summer seasons (**b**)

Three trees of the natural habitat in Bongae-dong are 10–11 m high and with a magnificent crown spanning 16 m in width. The flower of *P. yedoensis* blooms fully in the spring and the blossom enjoys such a long and splendid period that these trees are often selected to adorn parks and roads. A spectacular Cherry Blossom Festival is held every year in Jeju City (Fig. 6.10).

6.1.4.2 Natural Habitat of Cherry Trees (*Prunus yedonensis*) at Gwaneumsa Temple (Jeju Province Monument No. 51)

Located inside of Gwaneumsa, *Prunus yedoensis* tree are approximately 10 m high and 2 m around, having two trunks as its base. Although this is a rather small-sized tree it grows to be very healthy and presumably less than 50 years in age. *P. yedoensis* is a deciduous broad-leaved tree belonging to the rose family and grows nowhere else in the world except for its natural habitat of Hallasan, typically between 500–900 m above the sea level. The tree blossoms particularly well in the direct light of the sun (Fig. 6.11).

Fig. 6.11 Natural habitat of cherry trees (*Prunus yedoensis*) at Gwaneumsa temple

6.1.5 Legends

6.1.5.1 Baengnokdam crater lake

Once upon a time in Jeju there lived the 500 Generals, who were the sons of Grandmother Seolmundae, the mystical founder of Jeju. They mostly made a living hunting on Hallasan. One day when the hunt did not go very well, the eldest son vented his anger by firing arrows into the air. One of arrows struck the God of Heaven in the side. God of Heaven, pierced by an arrow, became so angry that he tore the rocky peak off the top of Hallasan and cast it away. From this, Baengnokdam crater lake formed in the vacant area that was created where the rocky peak was removed. The jettisoned rocky peak fell near the Sagyeri village and became Sanbangsan.

6.1.5.2 The 500 Generals in Yeongsil

There is a scenic area known as Yeongsil in the southwestern mountainside of Mt. Halla. It is surrounded by steep precipices and a thick forest. There are a large number of fantastic rocks and stones standing high to the sky, first interpreted to stand as Buddha's disciples and later seen as military generals. It is for this reason that they are called either the '500 Buddha's Disciples' or the '500 Generals'. The story of the 500 Generals is as follows:

Once upon a time there lived a mother with 500 sons. Because of her large family and a particularly lean year, she was having great difficulties making a living as so she asked her sons, "don't you think we have to do something to keep eating?" All 500 sons went out to get food and the mother began making a soup for their return. She placed a gigantic pot on the fire and stirred the soup by walking around the pot. Unfortunately, with a slight misstep, she fell into the pot. Soon after her sons returned home and began enjoying the soup, not knowing the fate of their mother. They all agreed that the soup tasted better than usual. The youngest son stirred the soup, before scooping his share, and discovered human bones within the broth. He then realized that their mother had melted into the soup. "I can stay no longer with such disrespectful brothers; who, with no respect for familial obligations, had a soup made from their own mother," deploring the misfortune and running away to Chagwido in Gosan-ri, Hangyeong-myeon, cried bitterly and was eventually transformed into a rock. Finally realizing the truth, the

Fig. 6.12 A number of pillar rocks in Yeongsil valley which is the site for the legend of '500 Generals'

other sons grieved endlessly. In the end, they were also transformed into rock structures which became known as the 500 Generals. There are, in fact, 499 generals in Yeongsil and the 500th general remains apart off on the island of Chagwido (Fig. 6.12).

6.1.5.3 Aheunahopgol (Ninety Nine Valleys)

There is a lava dome on the northern slope of Mt. Hallasan. It has such a variety of large and small-sized valleys, like furrows, that it is called 'Aheunahopgol', which means 99 valleys. It is believed that if there was one more valley in existence, the 100th valley, Jeju would be home of a large beast of prey such as a tiger or a lion. According to legend, there were originally 100 valleys in the mountain and many beasts of prey roamed wildly. One day a monk arrived in Jeju from China, gathered people and said that he will get rid of all wild beasts if they all shout out together, "the Great King of animals from China just arrived on the island." People did this with delight, and oddly enough, all beasts of prey gathered together at the mountain. After he chanted a sutra for some time, the monk shouted "Go away to the better place for you. The valley where you entered from will disappear. If you ever came back, all of your species will be destroyed!" Then all of the beasts, such as tigers, lions, bears, and so on, ran off into the valley which, as the monk stated, then disappeared and only 99 valleys were left. Since then no wild beasts of prey inhabited Jeju. Some people add that there were no kings or heroes that came from Jeju since there were no longer any beasts living there.

6.2 Seongsan Ilchulbong Tuff Cone Geosite

6.2.1 Geoheritage

Seongsan Ilchulbong, also called 'Sunrise Peak', is an archetypal tuff cone, 179 m high, dominating the eastern seaboard of Jeju Island like a gigantic ancient castle (Fig. 6.13). The

Fig. 6.13 An areal view of the Seongsan Ilchulbong tuff cone in the eastern margin of Jeju Island. Mt. Hallasan, the central peak of Jeju Island, can be seen in the background

tuff cone was produced by a hydrovolcanic eruption upon a shallow seabed about 5,000 years ago when the sea level was identical to that of the present (Sohn and Chough 1992; Sohn et al. 2002). Designated as a Natural Monument of Korea in 2000 and as a UNESCO's World Natural Heritage site in 2007, the tuff cone offers excellent scenery and attracts millions of tourists every year. In addition, the tuff cone reveals the birth and growth history of an emergent Surtseyan-type volcano erupted from a shallow seabed (Fig. 6.14). Virtually all kinds of deposit features that can be produced by Surtseyan-type hydrovolcanic eruptions are found in the tuff cone (Sohn and Chough 1992), not only offering information on past volcanic activity and depositional processes of the tuff cone but also providing a basis for interpreting eruptive/depositional processes of other phreatomagmatic and subaqueous volcanoes in other parts of the world.

6.2 Seongsan Ilchulbong Tuff Cone Geosite

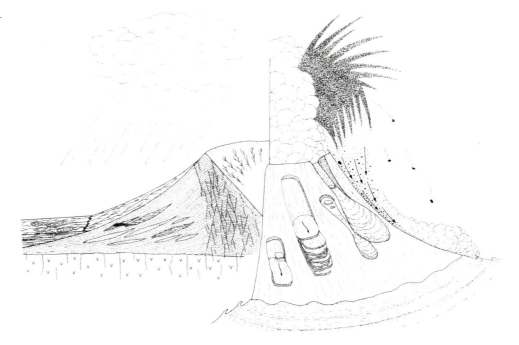

Fig. 6.14 Illustration of the eruption and depositional processes of the Seongsan Ilchulbong tuff cone. (after Sohn 1992)

Volcanic cones or "*oreums*" (Jeju dialect for volcanic cones) in Jeju Island are mostly scoria cones formed by Hawaiian or Strombolian eruptions. They are composed of dark-colored and vesicular volcanic rock fragments, called scoria. On the other hand, Seongsan Ilchulbong and several other oreums in Jeju Island are hydromagmatic volcanoes that were produced by explosive interaction of hot ascending magma and seawater or groundwater (Sohn 1996). Hydromagmatic volcanoes are classified into tuff rings and tuff cones based on crater size and height and the slope angle (Wohletz and Sheridan 1983; Vespermann and Schmincke 2000). Seongsan Ilchulbong has the typical morphology of a tuff cone with a height of 179 m, crater diameter of about 600 m, and the dip of strata up to 45°. The crater floor is 90 m above sea level.

Seongsan Ilchulbong is surrounded by precipitous cliffs except for the western flank because of erosion by marine waves (Fig. 6.15). The tuff cone therefore provides superb geological cross-sections of the volcano from the intracrater deposits to the marginal strata. Diverse geological structures are observable on the sea cliffs, including syn-depositional faults and fractures that formed during the eruption and accumulation of the tuff cone (Fig. 6.16), deposits of slide, slump, and debris flow that were generated by slope failure during the eruption (Fig. 6.17), armored lapilli produced by adhesion of fine and wet ash onto coarse-grained volcanic rock fragments (Fig. 6.18), adhesion ripples formed by wet pyroclastic density currents (Fig. 6.19), and thinly stratified tuff with diverse internal structures (Fig. 6.20). These structures indicate that abundant water could permeate into the volcanic vent of Seongsan Ilchulbong during its eruption and that the erupted volcanic materials were very wet and sticky (Sohn and Chough 1992; Sohn 1996).

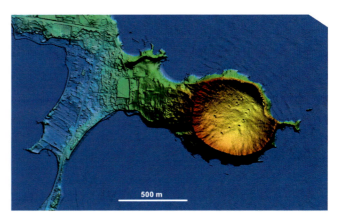

Fig. 6.15 Digital elevation model of Ilchulbong tuff cone obtained by LiDAR airborne terrain mapping. The crater rim and the crater floor of the middle-to-upper cone are 180 m and 90 m high, respectively

Fig. 6.16 Syndepositional faults and fractures near and inside the crater rim of the Seongsan Ilchulbong tuff cone

Fig. 6.17 Lens-shaped slide deposit with backset stratification

Fig. 6.20 Thinly stratified tuff with an internal erosion surface

Fig. 6.18 Armored lapilli produced by adhesion of fine and wet ash onto coarse-grained volcanic rock fragments

Fig. 6.19 Adhesion ripples formed by wet pyroclastic density currents

All these structures have great geological importance because they provide a basis for interpreting eruptive and depositional processes of hydromagmatic volcanoes worldwide in addition to the past volcanic activity of the Seongsan Ilchulbong tuff cone. There are numerous hydromagmatic volcanoes in the world similar to the Seongsan Ilchulbong tuff cone. However, Seongsan Ilchulbong is probably the only example of hydromagmatic volcanoes that has the typical morphology of a tuff cone and shows diverse internal structures along the sea cliff exposures.

6.2.2 Historical Heritage

6.2.2.1 Remains of Susan Jinseong (Jeju Province Monument No. 62)

Susan Jinseong is a part of military facilities called Gwanbang, a fortress to guard the border areas, built for the defense against the Japanese invaders during the Joseon Dynasty. During that period, Jeju had the Gwanbang facilities in

Fig. 6.21 Remains of the stone wall enclosing the fortress, Susan Jinseong during the Joseon Dynasty

Fig. 6.22 Remains of Japanese military caves on the sea-cliff of the Seongsan Ilchulbong tuff cone

Fig. 6.23 A view of Siksanbong scoria cone (**a**), which is a natural habitat for various low-land primitive flora including hamabo mallow (*Hibiscus hamabo*) (**b**)

12 locations of strategic importance, including 3 Eupseong (main bases) and nine local fortresses. Susan Jinseong as a local fortress was first constructed in 1439 for the purpose of watching for the warships of Japanese pirates in the Udo region. Its castle walls were built as a rectangular shape with 134 m width between the east and the west perimeter and with 138 m between the north and south areas. The total length is 544 m and stands as the fourth largest among fortresses. It says that there were two castle gates, a well, and other facilities such as offices, a guesthouse, and an arms storage building. Interior facilities no longer remain but the original formation of the castle walls is in reasonable condition compared to others fortresses in Jeju (Fig. 6.21).

6.2.2.2 Japanese Military Caves (Registered Cultural Heritage No. 311)

Seongsan, the eastern side of Jeju, is where the 58th Japanese Army headquarters built its partisan position under their plan of fortifying Jeju during the World War II. The Japanese marine corps. base was constructed at the seashore, which at that time was known as "Soomapo". There are still 21 cave dug-outs remaining, which the Japanese Army built through the mobilization of Korean laborers; most brought in from the coal mining areas of Hwasoon in Jeollanam Province, the southwestern Korean Peninsula. The average cave is 3–6 m high and about 30 m deep and was used to hide bombs and torpedoes or to house communication equipment. It is also said that the Japanese Army attempted to set up a radar station at the east end, inside of the Seongsan Ilchulbong crater. Its forces were stationed in the area where the parking lot is today. This Japanese naval base was designed to launch torpedo attacks against U.S. naval vessels (Fig. 6.22).

6.2.3 Biological Heritage

6.2.3.1 Siksanbong (Jeju Province Monument No. 62)

Siksanbong is a cinder cone of 66 m altitude at the seashore of north-west from Seongsan Ilchulbong. The name of Siksanbong represents a shape of a barley stack. The origin of its name is based on the historical reference that Siksanbong was fully covered with straw thatch to camouflage military provisions and to frighten off the Japanese invaders who frequently attacked the areas near Seongsan-ri. In ancient times, Siksanbong was the only area of low-land primitive flora, such as a warm-temperate evergreen forest and coastal plant, which once grew in the eastern areas of Jeju. In particular, about 20 hamabo mallow (*Hibiscus hamabo*) (Fig. 6.23), now designated as natural treasure, grow on these shores of high salinity. *Desmodium caudatum*, *Callicarpa japonica*, *Lygodim japonicum*, *Machilus thunbergii*, *Neolitsea sericea*, *Piper kadzura*, *Litsea japonica* also grow in this area.

6.2.3.2 Tidal Flats on Shiheung-ri and Jongdal-ri coasts

Vast intertidal flats, made up of high-quality fine sands, have developed in the coast of Shiheung-ri and Jongdal-ri

Fig. 6.24 Sandy tidal flat on Jongdal-ri coast

Fig. 6.25 Deunggyeongdol, a columnar rock on the slope of Seongsan Ilchulbong tuff cone

(Fig. 6.24). It is the only foreshore area of Jeju, where short-necked clams and razor clams can be harvested. There are shores of high salinity with reeds that spread deep into the inlands of Shiheung-ri and Jongdal-ri, creating a perfect habitat for what is now a migratory bird sanctuary over the winter season.

6.2.4 Legend of Deunggyeongdol

There are 99 sharp rocks standing in a circle on top of Seongsan Ilchulbong. One of them was being used to support an oil-lamp when Grandmother Seolmundae was sewing. The lamplight was too low for her so she raised the oil lamp by adding a second rock to the support. This rock is still called 'Deunggyeongdol', meaning a rock supporting the oil-lamp, as applied by Grandmother Seolmundae (Fig. 6.25).

6.3 Manjang Cave (lava tube cave) Geosite

6.3.1 Geoheritage

The Manjang Cave (Manjanggul in Korean) is a part of the Geomunoreum Lava Tube System and a single entity that meanders along the Geomunoreum lava flow. The cave is a very popular tourist attraction with hundreds of thousands of visitors annually. The Geomunoreum Lava Tube System consists of a number of lava tube caves (Bendwi, Utsanjeon, Manjang, Gymneong, Yongcheon and Dangcheomul caves). Stretching for as far as 7,416 m, its length ranks as 15th in the world. It is a large cave measuring up to 23 m in breadth and 30 m in height. The flow is of pahoehoe basaltic lava, mostly composed of alkaline periodotite basalt, the cave formed by multiple episodes of lava effusion in the Middle to Late Pleistocene.

Manjang Cave is a two-story cave; the lower level (the main tube) is 5,296 m long, and the upper level (a tributary) is 2,120 m long. Inside the cave, diverse structures and the innumerable cave formations, commonly found in lava tubes, are readily found. They include lava stalactites, lava stalagmites, lava columns, lava flowstone, lava helictites, lava foam, cave corals, lava roses, lava balls, bridges, shelves and striations all formed in the pahoehoe (ropy) lava. Although there are three entrances leading into the Manjang Cave, only Entrance Number 2 and a one kilometer length, heading southward inside the entrance, are open to the public. At the end of the showcave passage, a large amount of lava spilled from the upper level down to the floor of the lower level. Eventually, the lava united the floor of the lower level with the ceiling, creating a very large column standing 7.6 m high. The inside of the pillar is hollow, possibly suggesting that it once served as a gas release tube (Figs. 6.26, 6.27, 6.28, 6.29 and 6.30).

Fig. 6.26 Main passage of the lower level in Manjang Cave

6.3 Manjang Cave (lava tube cave) Geosite

Fig. 6.27 Lava bridge of the upper level between 1st and 2nd entrance of the Manjang Cave

Fig. 6.28 Lava Column at the end of the tourist route in Manjang Cave. It is known to be the largest lava column in the world

Fig. 6.29 Pahoehoe (ropy) lava beautifully preserved on the floor of the Manjang Cave

Fig. 6.30 Visitors in the tourist passage of the Manjang Cave

6.3.2 Historical Heritage: Remains of Hwanhae Great Wall in Haengwon and Handong (Jeju Province Monument No. 49-7 and 49-8)

A stone wall running for approximately 120 km, or nearly the entire shore of the island, is called "the Great Wall of Jeju". It was built during the time when the Sambyeolcho Army went to Jindo Island to fight against the Mongol Army, the Goryeo Court, who had surrendered to the Mongol Army, sent troops and ordered them to build the walls in order to prevent the Sambyeolcho Army from entering Jeju. The Sambyeolcho Army occupied Jeju in 1270 after defeating the Goryeo Army. Afterwards, the Sambyeolcho Army continued to build the walls to defend themselves against the Goryeo and Mongol Allied Armies. During the Joseon Dynasty, the walls were under constant maintenance, as a military protection against attacks from the Japanese pirates and foreign warships. Hwanhae Great Wall is made up of basalt rocks grouped by size. Its style varies and a total of ten specific locations, deemed in of worthy condition, were designated as portions of a Jeju

Fig. 6.31 Remains of Hwanhae Great Wall in Haengwon-ri (**a**) and Handong-ri (**b**) constructed for the defense of foreign enemies during the Joseon Dynasty

Fig. 6.32 Bijarim (**a**) is a natural habitat of nutmeg trees (**b**)

Fig. 6.33 Entrance (**a**) and inside (**b**) views of Gimnyeong Snake Cave

Province Monument. There are two locations of Hwanhae Great Wall from the Joseon Dynasty remaining on the shores of Haengwon-ri and Handong-ri, 310 m in length and 290 m respectively. While the walls in Haengwon-ri are between 1.5 and 2.0 m high, the walls in Handong-ri are 3.0 and 3.8 m high, twice as long as the former. The castle walls were built in such a way that natural rocks, found easily on the shores, were first stacked with infrequent trimming and then rubble was added in order to fill gaps (Fig. 6.31).

6.3.3 Biological Heritage: Bijarim (Natural Monument No. 374)

Bijarim is a netmeg (Torreya nucifera) forest covering 448,165 m^2, is located in almost completely flat portions of Gotjawal, where blocky aa lava or fragments of volcanic rocks are exposed along the ground (Fig. 6.32). The maximum age of the trees is in the area of 900 years and their average height is between 11–13 m and is of an evergreen tree belonging to the yew family. Its bark is grey-brown in color and its leaf is 25 mm long and 3 mm wide. The nutmeg tree has separate male and female plants. It blossoms in April and May, and bears fruit in September and October of the following year. The nutmeg nuts have been used as medicine, which was an important tribute to a past king. Its superb wood is known for use in quality furniture or padook boards. Rare orchid plants, such as *Aerides japonicum*, *Neofinetia falcata*, *Bulbophyllum*, *Liparis nervosa* Lindley and *Oberonia japonica* also grow in Bijarim.

6.3.4 Legend of Gimnyeongsagul (Gimnyeong Snake Cave)

Gimnyeongsagul is a cave that seems to have been once connected to the Manjanggul lava tube (Fig. 6.33). Once upon a time, there was a huge snake living in the cave and thus the cave was called 'Baemgul' meaning 'snake cave'. The snake was extremely big and it had a gigantic body approximately the size of a 240-gallon pot. People from the village had a yearly exorcism by sacrificing one girl. Without exorcising, the snake would have come out of the cave and trampled all the farming fields causing the village to suffer a year of famine. So every year one girl from the village had to be sacrificed. Usually daughters from the lowly were made a victim because noble families would seldom offer their daughters. From this, the daughters of the more lower-demographic families, like shamans, did not have high prospects for marriage.

About 500 years ago during the King Joongjong period in the Joseon Dynasty, a newly appointed judge arrived in Jeju Island. His name was Seo Ryeon and his age was 19 years old. Judge Seo was greatly irritated on hearing the story of Baemgul. He soon instructed the people to have an exorcism with not only the girl but also with a collection of foods. He then lead his soldiers off to Baemgul in Gimnyeong. While the exorcism was taking place, the gigantic snake came out and first devoured the food and drinks and then began moving toward the girl. At this moment Judge Seo and his soldiers attacked the snake and killed it with spears and swords. By observing the process, the shaman said to Judge Seo, "Please hurry your horse and go back to the castle in the city now. Never look back on the way from which you came." So Judge Seo whipped the horse on and returned to the castle. When he arrived at the east gate of the castle, one soldier shouted out, "Blood rain is falling behind us!" "Why on earth is there a blood rain?", when Judge Seo looked back involuntarily he immediately fell to the ground and died. The blood rain was believed to be the blood from the snake, which turned into rain to chase after and revenge Judge Seo.

6.4 Seogwipo Formation and Cheonjiyeon Waterfall geosites

6.4.1 Geoheritage

Thousands of groundwater bores have been drilled all over the island since the 1960s, revealing subsurface stratigraphy of Jeju Island (Fig. 6.3). The most remarkable finding was that the lavas of the island are only 40 m thick along the coastal regions and are underlain by *c.* 100 m of volcaniclastic deposits almost all around the island (Sohn and Park 2004; Sohn et al. 2008). The volcaniclastic deposits, named the Seoguipo Formation, are generally impermeable and act as an important aquiclude throughout Jeju Island, thus controlling significantly the behavior of groundwater flow (Koh 1997).

Fig. 6.34 Coastal cliff near the Seoguipo Harbor, showing exposures of the Seoguipo Formation

Fig. 6.35 Palm-size molluscan shells within the basaltic volcaniclastic deposit of the Seoguipo Formation

The Seoguipo Formation is exposed only in the south-central part of Jeju Island along a cliff wall overlain by a 400,000 year old hawaiite lava (Lee et al. 1988) (Fig. 6.34). The 40-m-thick exposure of the formation is composed of mainly basaltic volcaniclastic rocks and subordinate amounts of non-volcanic sedimentary layers. The lower half of the exposure is characterized by shallow-marine lithofacies and consists of alternations of fossiliferous and non-fossiliferous units, whereas the upper half is mostly nonmarine and non-fossiliferous. Overall, the formation comprises a deepening-to shallowing-upward megasequence with several transgressive and regressive cycles in it.

A variety of fossils are contained in the fossiliferous units of the exposure, including mollusks, brachiopods, foraminifera, ostracodes, nannofossils, sponges, corals, barnacles, echinoids, shark teeth, bryozoans, and whale bonds as well as prolific trace fossils (Yokoyama 1923; Haraguchi 1931; Kim 1972; Yoon 1988; Yi et al. 1998; Kang et al. 1999; Li et al. 1999; Kang 2003) (Fig. 6.35). Because of the fossil diversity, the formation was designated as a

Fig. 6.36 Logs of the Seoguipo Formation and the volcaniclastic Unit IX. **a** The Seoguipo Formation contains four volcaniclastic units (Units III, IX, XI and XIII) that are devoid of fossils and bioturbation. **b** Unit IX consists mainly of normally graded and ripple cross-laminated deposits, which we divided into 24 subunits. (after Sohn and Yoon 2010)

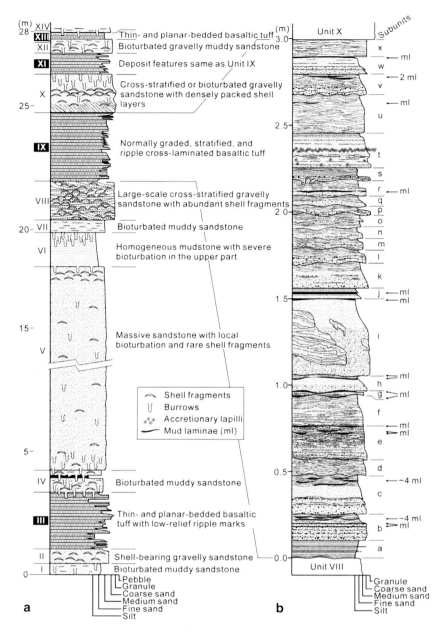

natural monument of Korea (no. 195) in 1968. These fossils suggest that the formation accumulated throughout the Early Pleistocene.

The outcrop section of the formation was studied in detail by Yoon and Chough (2006) and Sohn and Yoon (2010) (Fig. 6.36). The outcrop section can be subdivided into 10 fossil-bearing and 4 fossil-free units that are distinguished by either sharp erosional surfaces or distinct changes in lithology, fossil occurrence, grain size and sedimentary structures. Analyses of the sedimentary facies suggest interplay of a variety of depositional settings including storm-dominated shoreface with sporadic input of volcaniclastic materials, sandy nearshore to inner shelf, and mud-dominated outer shelf.

Recent study of a volcaniclastic unit with diverse current-induced sedimentary structures (Fig. 6.37) suggests that the materials of the unit were transported to the water surface by pyroclastic clouds and then settled from the surface as they were entrained in the water. The deposition is interpreted to have occurred under alternating currents and still waters, which is most plausibly attributed to tidal processes. Mud flasers or drapes intercalated in the deposit, which indicate periods of slackwater during tidal cycles, suggest that the deposit accumulated in a very short period of a fortnight or a month, about a million times faster than the adjacent sedimentary strata. Because of the unusually high sedimentation rate, the volcaniclastic deposit could record the 'usual' fair-weather processes in the depositional site at a resolution that

Fig. 6.37 Outcrop photographs from Unit IX of the Seoguipo Formation. **a–c** Normally graded and cross-laminated deposits formed by a combination of tidal currents and pyroclastic fallouts. **d–f** Parallel stratified deposits with accretionary lapilli formed during slackwater periods

is almost never provided by ordinary sedimentary deposits. This finding highlights the biases in Earth's stratigraphic records and teaches us that volcanic deposits, commonly regarded as the products of catastrophic events, can in some cases record more faithfully the ordinary and usual processes that non-volcanic deposits cannot.

6.4.2 Biological Heritage

6.4.2.1 *Anguilla marmorata* (Natural Monument No. 27)

Anguilla marmorata is a large tropical eel, belonging to the family Anguillidae. *A. marmorata* is widely distributed over

Fig. 6.38 Cheonjiyeon waterfall and the warm-temperate forests along the downstream valley

Fig. 6.39 Photos of Munseom (**a**) and Beomseom (**b**)

Fig. 6.40 Remains of Japanese military caves on the sea-cliff of Hwangwooji coast

Fig. 6.41 Jeongbang waterfall on the 25 m high trachy-andesite cliff

Fig. 6.42 Photos of the Oedolgae sea stack

6.4 Seogwipo Formation and Cheonjiyeon Waterfall geosites

Fig. 6.43 An overview of Beomseom (**a**) and a close-up view of sea caves (**b**)

the eastern area of Africa, the South Pacific, Southeast Asia, Japan and Taiwan. It is infrequently found in Tamjingang, Seomjingang, Geojedo and Yeongdukosipcheon, and from time to time it is found in the pond of Cheonjiyeon Waterfall. Because it is such rare fish, the entire species has been protected as a natural monument. Its northern boundary is Jeju, along with Nagasaki in Japan. *A. marmorata* lives in fresh water for 5–8 years and then lays eggs out in the deep sea.

6.4.2.2 Warm-Temperate Forests near Cheonjiyeon Waterfall (Natural Monument No. 379)

The Cheonjiyeon Valley contains various vascular plants, a total number of 447 species (Fig. 6.38). Seventeen species of Korean plants grow in only Jeju including the endemic species, *Rubus hongnoensis*. Warm Temperate Forests near the Cheonjiyeon Waterfall are one of the most famous evergreen forests in Korea. Designated as a natural monument, *Psilotum nudum* grows on a rock cliff. Broad-leaved evergreen trees like *Korthalsella japonica*, *Camellia japonica*, *Litsea japonica*, *Machilus thunbergii*, *Neolitsea sericea* and *Distylium racemosum*, various evergreen vines like *Ardisia crenata* Sims, *Eurya japonica*, *Piper kadzura*, *Ficus nipponica*, *Elaeagnus macrophylla*, *Elaegnus glabra*, *Hedera rhombea* and *Trachelospermum asiaticum var.* intermedium, and many ferny plants like *Lastrea subochthodes* and *Diplazium virescens*, all grow in the area.

6.4.2.3 Munseom and Beomseom Nature Reserve (Natural Monument No. 421)

About 1.5 km from the Seogwipo Coast, Munseom and Beomseom Nature Reserves include Munseom, designated as a natural monument on July 18th, 2000, Jaggeunmumseom, or Uitaldo, Beomseom, and Jaggeunbeomseom, or Choodo. Munseom and Beomseom are composed not from basalt but rather from trachyte. With well-developed rock formations and the presence of beautiful caves, formed by the seawater erosion, this area makes for picturesque scenery (Fig. 6.39). Rare evergreen broad-leaved trees such as *Korthalsella japonica* grow in the area. Munseom and Beomseom Nature Reserve is also the southern boundary of *Columbia janthian* which is designated as Natural Monument No. 215. Along the seashore, there are numerous marine creatures of great academic values. Typical Korean sea creatures, such as *Hypoglossum simulans*, and many non-registered species inhabit in this area. The Munseom and Beomseom Nature Reserves are considered among the most highly-valued areas as they represent the great diversity of southern species.

6.4.3 Historical Heritage: Sammaebong Japanese Military Caves

Hwangwooji sea cliff of hyaline tuff is located west of Seogwipo Formation outcrop. At the end of the Pacific War, the Japanese Army built cave dug-out positions in this area. The caves are thus referred to by the names of "Hwangwoojigul", or "Yeoldoogul". Twelve of them remain along this scenic seashore (Fig. 6.40).

6.4.4 Legends

6.4.4.1 Dragon of Cheonjiyeon Waterfall

In the middle of the Joseon Dynasty, in the town of Seogwipo, there was a sweet tempered and good natured girl named Suncheon. There was also a young man, Myeongmun, who was in love with her. When Suncheon got married to a man from the Gang family in Beophwan-ri, as her parents wanted, Myeongmun was heartbroken and his mind started wandering. Meanwhile, Suncheon was in the living very happily in her marriage, being respectful to her husband and in-laws, and winning high praise and a good reputation. One day when she was on the way to visit her parents, Myeongmun was aware of her travels and waited for her at the entrance of Cheonjiyeon. When Suncheon passed by, he grabbed her hands with a longing look and tried to hold her. "My dear Suncheon, I can't live without you. I will jump into Cheon-jiyeon Waterfall and die if anybody stops me from having you!", he cried out. In a moment, Suncheon screamed desperately, "Help!", and tried to run away. Then a dragon came out of a pond, snatched Myeongmun, and flew into the sky. When thanking the dragon for saving her, Suncheon found Cintamani, the "as-one-wishes" jewel, shining and rolling near the pond. Out of delight, she ran back to home holding Cintamani in her hand. She secretly kept Cintamani and all with her family was prosperous and happy. People among her family attributed their success to Suncheon's virtue.

6.4.4.2 Jeongbang Waterfall

Jeongbang Waterfall (Natural Monument No. 43) on the coast of Seogwipo City is one of the most well-known waterfalls in Jeju Island. The waterfall rises from a riverbed springwater 400 m from the coast. It runs down through the 25 m high trachy-andesite cliff with pillar-shaped joint and flows into the seawater, forming 5–7 waterfalls, 10 m width. They are a spectacular sight in harmony with scenery sea cave and escarpment (Fig. 6.41). According to Legend, Seobul, or Seobok, came to Jeju in search of an elixir of eternal youth for Chin's first emperor of China. Unable to accomplish his objective, before returning to the west, he inscribed "Seobulgwacha", meaning Seobul had passed here name Seogwipo. Based on the legend, the Seobul Exhibition Hall displays related to the history of Seogwipo and the Chinese emperor of Chin.

6.4.4.3 Oedolgae

Located in the west 1.2 km from the entry of Seogwipo Formation outcrop, Oedolgae is of sea stack, a leftover hard section along the coast, which has survived the longstanding process of wave erosion (Fig. 6.42). Oedolgae is 23 m high and 7–10 m wide. There is a centripetal-shaped joint that has developed at the bottom and small bushes at the top. Oedolgae refers to the fact that it stands alone in the middle of the sea. It is also called Janggunseok: during the end of Goryeo Dynasty in 1374 when the Goryeo Court often mobilized Jeju horses to offer to China in the Ming Dynasty, Jeju Mokho, the Mongolian shepherds, rebelled against the government. The king sent General Choi Yeong to Jeju to suppress a rebellion. While in pursuit during a final battle, Mokho fled onto Beomseom where Oedolgae is now visible off in the distance. General Choi created Oedolgae to represent the figure of a General and it is said that, on seeing this massive figure, Mokho committed suicide out of sheer terror.

6.4.4.4 Beomseom (Tiger Island)

Once upon a time, a hunter touched the belly of the God of Heaven. Out of his rage, God of Heaven tore out the hills of Mt. Halla and cast them away. One of them became Beom-

seom. It is believed that one sunny day, around noon, a black dragon ascended to heaven from the southeast of Beomseom. The head and body of the dragon were seen, through deep-black clouds, wriggling and soaring up into the sky.

There are two holes due north of Beomseom. Because they look like the nostrils of a tiger, these holes are called 'Kotkyeum' or the nostril caves (Fig. 6.43). Beomseom is facing directly toward Baengnokdam crater lake and Yeongsil so that Beomseom is also related to the legend of Grandmother Seolmundae. When she slept, Grandmother Seolmundae laid her head on Baengnokdam as a pillow, put her back on Gogeunsan, and stretched her feet to Beomseom. Kotkyeum were the holes where her toes reached.

There is a cave (Keunhangmunido), large enough to hide about 30 boats within its walls. According to legend, someone entered the cave by a small boat to find a very beautiful light glowing in the darkness. He then found a tree blossoming on the cliff, picked its flowers, and brought them home. The flowers were uniquely gorgeous, rarely seen anywhere else. He attempted to go back to the same location in the cave but couldn't find it anymore. People say that the light may have come from the shining pieces of axes, used by the Mongolian shepherds (Mokho) who ran into the cave after losing a battle.

6.5 Jisagae Columnar-Jointed Lava Geosite

6.5.1 Geoheritage

Volcanic rocks formed by cooling of hot lavas commonly have cooling joints, which are typically expressed as vertical columnar jointing. The best exposures of columnar-jointed lavas in Jeju Island are found along the coast of Daepodong where dark gray trachybasalt lava crops out for about 3 km (Fig. 6.44). The joint systems in this area are mostly six-sided, but there are also some four-sided to seven-sided joints (Koh et al. 2005). In addition to the polygonal jointing in plan, the columnar-jointed lava of Daepodong shows well-developed colonnade (the lower zone of columnar jointing that has thicker and better-formed columns) and entablature (the upper zone of columnar jointing that has thinner and less regular columns) in vertical section. The columns of the lava are locally curved and inclined, indicating that the lava had a lobate geometry (Fig. 6.45).

6.5.2 Cultural Heritage

6.5.2.1 Nonjitmul, a Coastal spring

Although Jeju is a rain-rich region with more than 1,500 mm of precipitation per year, most of the rainwater soaks into

6.5 Jisagae Columnar-Jointed Lava Geosite

Fig. 6.44 Columnar-jointed lava at Daepodong coast, showing well-developed colonnade and entablature

Fig. 6.45 Inclined columns of the Jisagae columnar-jointed lava

the underground and quickly runs through the layers of lava bed due to the large presence of rock and, in particular, a soil fabric of very high water permeability. The groundwater then rises to the ground surface in areas near the seashores due to the dense underground seawater. In the past, this presence of spring water in the coastal areas, called Yongcheonsu, led most residential areas in Jeju to develop near the seashore. It was routine to transfer water from the coast to residences by using Heobeok (pots) and Mulgudeok (bamboo baskets), and it was common to help by sharing water with neighbors who had an important family affair such as weddings, funerals or house construction. With modern water-work distribution over the entire region, the use of Yongcheonsu has decreased. By taking advantage of the existence of the great volume of water, a freshwater swimming pool was built in Nonjitmul in Yeye-dong (Fig. 6.46). Tourists are attracted to this area, whose presence is also aided by the awareness created by a local festival.

6.5.2.2 Irrigation Canal (Registered Cultural Heritage No. 156)

Because Jeju is a volcanic island with an abundance of underground water, rice has been a priceless crop, produced only in a few villages, since ancient times. In order to make rice farming possible, Jungmun residents led by Chae Guseok, Yi Jaeha, and Yi Taeok, built Cheonjeyeon irrigation canal system (Fig. 6.47), which takes the water from the first waterfall of Cheonjeyeon onto farm fields in front of Beritne Oreum, or Seongcheonbong, near Jeju International Convention Center and Jeju International Peace Center. The first construction work started in 1905 and finished in 1908. Breaking passageways through extremely dense rock bed was accomplished by heating with wood-piled fires and potent alcohols. Approximately 16.5 ha of rice fields were cultivated. The second construction phase took place between 1917 and 1923, creating an additional 6.6 ha of rice fields. Since then Jungmun Village, along with Gangjeong Village, has become the largest rice-farming area in Jeju.

Fig. 6.46 Nonjitmul, a spring on the seashore

Fig. 6.47 A part of Cheonjeyeon irrigation canal system built in the early 20th century

guarding against continuous attacks due to its geographical positioning as an island. Historical records show approximately 38 Yeondae locations in Jeju during the Joseon Dynasty. Smoke was used for daytime signals and flames were visible during the nighttime. A group of 12–36 workers took turns, working three shifts over a 24-hour period. Located on the sea cliff in the east, about 650 m away from the lookout of the Jisatgae columnar joints, is the Daepoyeondae beacon which was restored in 2000 (Fig. 6.48). Belonging to Daejeong-hyeon, it is 4 m high and 9.1 by 8.7 m square. It is built with basaltic stones.

6.5.3 Historical Heritage: Daepoyeondae (Jeju Province Monument No. 23-12)

Before the introduction of more advanced communication systems, a series of beacons, known as Yeondae, was set up on a number of hills near the ocean. A system borrowed from ancient times, Yeondae was strategically set in open coastal locations to aid in observing approaching enemy ships and

6.5.4 Biological Heritage: Warm-temperate Forests near Cheonjeyeon Waterfalls (Natural Monument No. 378)

It is a natural evergreen forest located on the both sides of the Cheonjeyeon valley (Fig. 6.49). A rare species, *Psilotum nudum* grows in the forest. The forest has excellent varieties of subtropical plants such as evergreen trees, deciduous trees, tendril plants, shrubs, and fernery. Researchers discovered a total of 363 classes of vascular plants, such as *Polystichum lepidocaulon*, *Sphenomeris chusana*, *Ophiopogon jaburan*, *Ficus nipponica*, *Impatiens textori* and *Ardisia pusilla*. On

Fig. 6.48 A restored beacon mound (Yeondae) near Daepo-dong, Seowipo city

Fig. 6.49 Cheonjeyeon waterfall (**a**) and the evergreen forests along the downstream valley (**b**)

6.6 Sanbangsan Geosite

Fig. 6.50 395-m high Sanbangsan Lava Dome at the southwestern margin of Jeju Island

Fig. 6.51 Vertical columnar joints on the southern cliff wall of the Sanbangsan lava dome. The lava dome contains a Buddhist temple on the middle of its southern slope which is a tourist destination in itself

the slopes of the waterfall, there is a perfectly preserved colony of *Castanopsis cuspidata* var. *sieboldii*.

6.6 Sanbangsan Geosite

6.6.1 Geoheritage

6.6.1.1 Sanbangsan Lava Dome

Sanbangsan is a gigantic lava dome located in the southwestern margin of Jeju Island, rising 395 m above sea level (Fig. 6.50). The lava dome resulted from the slow effusion of felsic and very viscous lava from a volcanic vent. The viscosity of the lava prevented it from flowing far from the vent, causing it to solidify quickly and creating a circular dome-like shape. Sanbangsan lava dome is composed of columnar-jointed trachyte lava, which is light gray in color and has a porphyritic texture with sanidine phenocrysts. The lava dome is one of the oldest rock formations on Jeju Island, being about 800,000 years old (Won et al. 1986). Vertical columnar joints, about 2 m in width and more than 100 m in height, are well exposed on the southern cliff wall of the lava dome (Fig. 6.51). Beneath the jointed lava occurs a thick layer of volcanic breccia composed of angular fragments of trachyte lava (Fig. 6.52). The breccia suggests that the dome experienced explosive eruptions or dome collapse during its growth due to build-up of gas pressure. The dome-forming lava and breccia overlies the basaltic tuff of Yongmeori (see below), the oldest rock formation on Jeju Island (Sohn 1995). A roadcut between the lava dome and Yongmeori shows clearly the relationship between these two rock formations (Fig. 6.52).

6.6.1.2 Yongmeori Tuff Ring

Yongmeori, meaning 'dragonhead', forms a small promontory to the south of the Sanbangsan lava dome (Fig. 6.53).

Fig. 6.52 Roadcut exposure of volcanic breccia beneath the Sanbangsan lava dome

Fig. 6.53 A panoramic view of the Yongmeori tuff ring with the Sanbangsan lava dome in the background

The overall geologic structures suggest that Yongmeori is a remnant of a tuff ring, which is a kind of hydromagmatic volcano with a relatively large crater compared with its height (White and Houghton 2000). The overall thin stratification with undulatory or megaripple bedforms (Fig. 6.54) suggests that it was deposited mostly by pyroclastic surges (turbulent, ground-hugging flows of volcanic gases, steam, and pyroc-

Fig. 6.54 The Yongmeori tuff ring is composed of overall thinly stratified tuff with undulatory or megaripple bedforms, suggesting deposition from pyroclastic surges

lastic materials) (Sohn 1995). In addition, lenticular (either concave-up or convex-up) massive deposits are commonly intercalated, suggesting occasional generation of debris flows by either remobilization of wet tephra or expulsion of vent-filling tephra slurry (Sohn 1995). Overall, the eruption of Yongmeori is interpreted to have been very wet.

The pyroclastic deposits of Yongmeori show variable bed attitudes and lithofacies distribution (Fig. 6.55). The quaquaversal bedding in the southernmost part (Fig. 6.56) suggests that Yongmeori comprises part of the crater rim of a tuff ring, but the overall lithofacies characteristics are not explained by the proximal-distal facies relationships of a simple (circular) tuff ring. Careful observations of the complex reveal that it consists of three stratal packages that are bounded by laterally persistent truncation surfaces and originated from different source vents (Sohn and Park 2005) (Fig. 6.57).

Based on these observations, Sohn and Park (2005) concluded that Yongmeori resulted from superposition of multiple tuff-ring deposits that have contrasting bed attitudes and paleoflow directions. The laterally persistent truncation surface was interpreted to have formed during a break in eruptive activity after large-scale collapse of the substrate. Collapse was probably caused by the instability of the friable sedimentary substrate (the U Formation), removal of lateral support because of downward quarrying of volcanic conduits, and liquefaction of the water-saturated substrate by volcanic seismicity. The path of magma supply was probably diverted in some cases after collapse, giving rise to migration of the active vent. The resultant volcanic edifice thus became non-circular or irregular. All these processes were possible because Yongmeori was built upon unconsolidated shelf se-

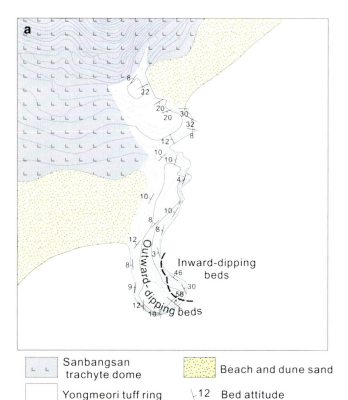

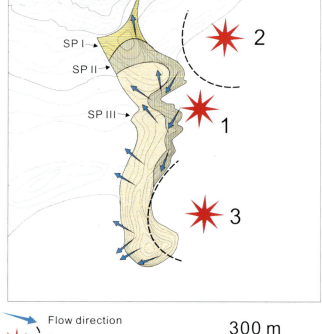

Fig. 6.55 Geology and morphology of the Yongmeori tuff ring composed of superposed multiple rim deposits or stratal packages (SP) that were derived from three different source vents (after Sohn and Park 2005)

6.6 Sanbangsan Geosite

Fig. 6.56 Quaquaversal bedding in the southernmost part of Yongmeori, suggesting it comprises part of the crater rim of a tuff ring

diment named the U Formation before the effusion of the shield-forming lavas of Jeju Island.

Yongmeori is also a historic site where a Dutch merchant ship was shipwrecked in 1653. One of the sailors in that ship introduced Korea to the western world for the first time via publishing a book on Korea after his 13-year-long detention in Korea.

6.6.2 Cultural Heritage

6.6.2.1 Sanbanggulsa (Sanbang Cave Temple)

Sanbanggulsa is a temple built within a cave in the southwest of Sanbangsan, which is at an altitude of 200 m (Fig. 6.55). There are no manmade temple buildings due to its natural cave location. It measures 10 m in length, is 5 m high and 5 m wide. According to the documents, it was a place where Priest Hyeil practiced Buddhist meditation and entered nirvana. The stone statue of Buddha placed during that time was taken by Japanese people under the Japanese rule. A seated stone statue of Buddha was enshrined again around 1960 and three contemporary temple buildings were constructed in the entrance in approximately 1962. Monks from the temple and residents of Sagye-ri work together today in order to maintain Sanbanggulsa. As one of the ten Scenic Wonders of Jeju, the beauty of its view during sun-

Fig. 6.57 Yongmeori consists of three stratal packages (SP) that are bounded by laterally persistent truncation surfaces and originated from different source vents (Sohn and Park 2005)

Fig. 6.58 Sanbanggulsa, a temple built within a cave on the slope face of Sanbangsan lava dome

Fig. 6.59 An overview of Daejeong Hyanggyo, a local educational institution in the Joseon Dynasty

rise defies description. One can witness the new day rising over the Yongmeori Cliff, Hyeongjaeseom, Gapado, and Marado along with the ancient pine trees standing outside of the cave. The interior of the cave is wrapped with rock walls and crystal clear water drops continually fall from the rock ceiling. The water is said to be an expression of tears of love from Sanbangduk, a goddess who guards the rock walls of Sanbangsan (Figs. 6.58).

6.6.2.2 Daejeong Hyanggyo (Jeju Province Tangible Cultural Property No. 4)

Located in the southern side of Dansanbong, west of Sanbangsan, Daejeong Hyanggyo (Fig. 6.59), is a local educational institution from the Joseon Dynasty. Since it was built inside the Daejeong castle in 1416 (the 16th year of King Taejong), it was relocated on two occasions before settling in its current location in 1653 (the 4th year of Hyojong). Daeseongjeon (main building) was renovated five times afterward, until 1993, and Myeongryundang was enlarged in 1772 (the 48th year of King Yeongjo). The architecture holds a five-room Daeseongjeon, Myeongryundang, Dongjae (east wing of the building) and Seojae (west wing of the building) in the precincts. Among other Hyanggyos in Jeju, Daejeong Hyanggyo is superior in maintaining its original condition. While living in exile, Chusa Kim Jeong-hee produced a calligraphic writing "Uimoondang" for its Myeongryundang.

Fig. 6.60 Remains of stone wall enclosing the fortress, Daejeong Castle in the Joseon Dynasty

6.6 Sanbangsan Geosite

Fig. 6.61 Hamel monument in front of the Sanbangsan lava dome (**a**) and a replica of the Hamel's sailing vessel (**b**)

Fig. 6.62 Restored exile site (**a**) and exhibition center (**b**) of Chusa Kim Jeong-hee

Fig. 6.63 Remains of Japanese military caves on the sea-cliff of the Songaksan

6.6.3 Historical Heritage

6.6.3.1 Remains of Daejeong Castle (Jeju Province Monument No. 12)

Located approximately 3.5 km northwest of Sanbangsan, the Daejeong Castle Site (Fig. 6.60), is a site where the castle of Daejeong-hyeon, one of 3 Eupseong (stronghold towns) in Jeju Island during the Joseon Dynasty, existed. In 1416 (the 16th year of King Taejong), Jeju was divided into 3 administrative districts such as Jeju-mok, Jeongui-hyeon and Daejeong-hyeon. Two years later, in 1418 (the 18th year of King Taejong), the first mayor of Daejeong-hyeon, Yushin, built the Daejeong castle. According to records, the castle

Fig. 6.64 Remains of Japanese military caves in the outer rim of the Songaksan

Fig. 6.65 Remains of aircraft hangars in Alteureu Airfield site, built and operated by Japanese Naval Air Force during the WWII

Fig. 6.66 A view of Yongmeori (Dragon Head) from the southern slope of Sanbangsan lava dome

was 1.5 km long and 5 m high. It is said that there were 4 gates; one to the north, south, east and west, respectively. The north gate was later closed permanently. It is assumed that there was a moat around castle walls. During that time there were about 540 troops stationed with Dongheon (main office) and other facilities within the castle.

It is also assumed that Dolhareubang (stone statue, Jeju Province Folklore Material No. 2), was the chief gatekeeper, with four Dolhareubang standing guard in front of each of the gates. Moved from the original locations, they were scattered to neighboring villages. Dolhareubang from the Daejeong castle is 108–146 cm high, smaller than usual, with the appearance reflecting a smiling face.

6.6.3.2 Hamel Monument

Hendrick Hamel and his company boarded the trading vessel Sperwer, under orders of the Dutch governor-general and the council, and departed Batavia, on June 18th, 1653 (the 4th year of King Hyojong). The vessel was transporting a new governor-general, and goods to Taiwan. Arriving in on July 16th, after a favorable voyage, Governor-general disembarked and the goods were delivered. Hamel and his company then departed for Nagasaki, Japan under the orders of the Taiwanese governor-general and council. Unfortunately the vessel was destroyed by a severe storm and only 36 of 64 crew members survived after drifting to the beach of Daejeong-hyeon in Jeju. They were sent to Seoul, detained for 2 years, and transferred to Jeollado in March 1656. Fourteen people died between 1656 and 1663. Twenty-two survivors were admitted separately to camps over Suncheon, Namwon, Yeosu and other places in the mainland of Korea. On September 4th, 1666 (the 7th year of King Hyeonjong), Hamel and 8 people escaped by sailing from the Jwasuyeong of Yeosu to Nagasaki, Japan. Hamel and his company stayed in Japan for a time and then returned to Holland in July 1668 after their 13 year detention in Joseon. In 1980, Korea International Culture Association and the Dutch Embassy in Korea erected the Hamel monument, on the hill in front of Sanbangsan, honoring Hamel and the mutual friendship between Holland and Korea (Fig. 6.61).

6.6.3.3 Chusa Exile Site (Historical Remains No. 487)

Located near the east gate of Daejeong Castle Site, Chusa Exile Site is where Chusa Kim Jeong-hee (1786–1856) lived (Fig. 6.62). He was a well-known calligrapher from the late Joseon Dynasty and renowned scholar for bibliographical studies, epigraphy, Buddhist studies and studies in Chinese classics. While he passed the state examination and obtained a government post as an academic official at Sungkyunkwan and a vice-minister, Chusa lost his political position due to a power struggle and was eventually exiled to Jeju. Over a period of 8 years he developed the Chusa calligraphy style, completing many paintings and writings including Seohando (National Treasure No. 180), and teaching classics and calligraphy to local Jeju students of Confucianism. Honored as the place where Chusa achieved his knowledge and art, the Chusa Exile Site was designated a historical location in 2007. The original house where Chusa had lived was destroyed by fire in 1948, and restored in 1984. Near the site, the Chusa Exhibition Center is under construction by Jeju Special Self-governing Province.

6.6 Sanbangsan Geosite

Fig. 6.67 Hyeongjeseom, a small islet consisting of two rocky peaks

6.6.3.4 Japanese Military Caves on the coast of Songaksan (Registered Cultural Heritage No. 313)

The Japanese positioned a marine base on the coast near Songaksan to have a strategic vantage point over the entire port area from Songaksan through to Hwasun. Along with Songaksan there were also artillery battery positions, bunkers and tunnels at Dansan, Sanbangsan and Wallabong. There are now ten tunnels remaining on the Songaksan coast (Fig. 6.63). Most are 3–4 m wide and 20 m deep and were used to conceal torpedoes in the same manner as the Japanese marine corps base at Seongsan Ilchulbong. Berth facilities for mooring suicide-boats and launching torpedoes were also constructed in this area.

6.6.3.5 Japanese Military Caves in the outer rim of Songaksan (Registered Cultural Heritage No. 317)

In and around Songaksan tuff ring, there were 14 tunnels constructed to face the direction of the ocean and 5 toward Al Oreum. Designed for potential use as an underground tactical position, the Japanese Army used these tunnels for the storage of war supplies and as air-raid shelters. In particular, the tunnels dug near Al Oreum were large enough to house military trucks transporting war supplies (Fig. 6.64). The numerous tunnels in Al Oreum were connected to one another to create a functionable underground fortress.

6.6.3.6 Underground Bunker at the Alteureu Airfield (Registered Cultural Heritage No. 312)

The Japanese Naval Air Force built its airfield in the area of the Alteureu plain as its topography made for a highly suitable landing strip. As the Sino-Japanese War broke out, its location was ideal as an advanced base for an attack on China. The first construction of the Alteureu airfield was planned in 1926 and took about ten years to complete. The Japanese Army began a second round of construction in 1937, and finished it in 1945 so as to enlarge the airfield from 66 to 264 ha. Residents from Aloreum-dong, Jeogeungae, Golmot and Gwangdaewon were evacuated from the construction zone. There are about 20 aircraft hangers remaining in the coastal direction of Alteureu Airfield (Fig. 6.65). Most remain intact, with the exception of two or three which have sunk slightly. These aircraft hangers were built to protect the Japanese suicide-bombers, so called 'Kamikaze Pilots' from air raid. The underground bunker, located in the northeast of the airfield, still maintains its the original condition. Beneath Alteureu airfield there is a two-floor structure with power generators believed to be used as communication facilities.

6.6.4 Legend

6.6.4.1 Sanbangsan

Once upon a time in Jeju there lived the 500 Generals, who were the sons of Grandmother Seolmundae, the mystical founder of Jeju. They mostly made their living from hunting in Mt. Halla. One day when the hunt did not go very well, the eldest son vented his anger by firing arrows into the air. One of the more sharp-tipped arrows hit God of Heaven directly in the seat of the pants. God of Heaven became so angry that he tore the rocky peak off the top of Hallasan and cast it away. From this, Baengnokdam crater lake formed in the vacant area that was created where the rocky peak was removed. The jettisoned rocky peak fell near the Sagyeri village and became Sanbangsan.

According to another legend, Sanbangsan used to be a peak of Hallasan. Once upon a time, one hunter went on a deer hunt. While wandering on Hallasan without a success, he arrived at the top of the mountain. Hallasan was so high that it seemed to reach the sky. The hunter finally found one deer and started chasing it holding a bow high. By mistake, he hit God of Heaven's bottom with the head of the arrow. In a rage, God of Heaven pulled out the peak of Hallasan and threw it out toward west. The peak landed and became Sanbangsan, and the dug spot on the top of Hallasan became Baengnokdam crater lake.

6.6.4.2 Sanbanggulsa

Sanbanggulsa is a temple built within a natural cave (named Sanbanggul) of Sanbangsan lava dome. Water drops falling from the center of the ceiling of Sanbanggulsa is a curious mystery. There is only a small amount of water that drips within a given year and yet the taste is extraordinarily good. Although there are often waters flowing between the rocks, rarely does water ever fall from the ceiling of the rocky cave, let alone on a continual basis. This crystal water is called Sanbangdeok's tears.

Sanbangdeok, a goddess of the rocky cave, missed the human inhabitants and prayed for 100 days for God to make her human. On the night of 100th day, in a thunderstorm, Sanbangdeok became human and, among a flurry of lightning, came down to earth. She was a lady of great beauty and refined manners. She found a man named Goh Seong-mok, hardworking and a good hunter, and decided to be his mistress. Goh Seong-mok built an orchard and a house near 'Gonmul', famous for its tasty waters. In the heavy rainy season, he hunted hundreds of wild boars, made a tent of their skins, and then pitched the tent around Sanbangdeok's house. The minister of Jeju heard of the story of Sanbangdeok and became greedy for her beauty so he had Goh Seong-mok arrested under false charges of building a large house beyond his social position as a commoner. After Goh Seong-mok was sent to jail, Sanbangdeok went to the minister and beg him to release her innocent husband. Since he initially planned to take Sanbangdeok away from her husband, the minister didn't listen to her appeal. In the end Goh Seong-mok didn't survive and his house and property were completely destroyed. Deploring the evil and wicked reality of human existence Sanbangdeok finally decided to leave. However, the greedy minister sent officials to kidnap her. At that moment, a white horse suddenly appeared in front of Sanbangdeok and took her into the cave of Sanbanggul. Officials chased her to the entrance of the cave, but she was already gone. Inside the cave, they found only a faint shape of the white horse. It was believed that Sanbangdeok returned to the cave and become a rock in order to flee the vices of the human world. From this, the crystal clear water began dripping from the roof of the cave and is believed to be Sanbangdeok's tears. Visitors take only three sips of the water, believing that three sips bring them a blessing, where taking more would reveal them as greedy.

6.6.4.3 Yongmeori (Dragon Head)

In Yongmeori, there is the legend of Goh Jong-dal who is from the period of Chin's first emperor of China. The Chinese emperor of Chin protected his reign by not allowing any potential emperors to be born in neighboring countries. One day he heard the rumor that a potential emperor would be born in Jeju at the specific area where lords are born known as Sanbangsan. Soon Chin's first emperor of China sent Goh Jong-dal, who was an expert in the feng shui theory. While he was investigating the entire area of Sanbangsan, Goh Jong-dal found Yongmeori shaped like a dragon (Fig. 6.66). He cut off its tail with a single stroke of a sword and cut its back with two strokes. The blood ran down the rocks and Sanbangsan cried out with a groan. Because of this, there was no king born in Jeju.

6.6.4.4 Hyeongjeseom (Brother Islet)

Hyeongjeseom is an uninhabited islet located in the sea near Sanbangsan. The islet is mainly made of two large rocks, with other rocks on the seashore and small black stones which appear at low-tide (Fig. 6.67). Hyeongjeseom legend states that, based on the shapes of two main rocks, once upon a time two dragons exchanged formal bows to one other. Basically, depending on the direction one is viewing the two rocks, the appear very different; the shape of the islet looks like two dragons joined together when low-tide exposes the lower area of the islet.

Additionally, a detailed record of August, 1712, states that two dragons fought in the sea at Hyeongjeseom. It caused a tsunami and typhoon, and seriously damaged the village on the other side of the islet. The record shows that 66 houses were destroyed, a pine tree forest on the seashore was devastated, and the sand blown by storm covered all farm fields.

6.7 Suweolbong Geosite

6.7.1 Geoheritage

Suweolbong is a low-relief mount of pyroclastic deposit located at the western margin of Jeju Island (Fig. 6.68, 6.69). The maximum thickness of the deposit is about 70 m. Present topographic contours of the mount together with a sedimentological study (Sohn and Chough 1989) suggests that the mount represents the partly preserved rim beds of a tuff ring, whose vent lies several 100 m seaward of the present shoreline.

The pyroclastic deposit of Suweolbong provides excellent and continuous sea-cliff exposures (Fig. 6.70). The exposures of the Suwolbong tuff ring are so continuous that individual tuff layers can be correlated bed-by-bed from the proximal to the distal parts of the entire tuff ring deposit (Fig. 6.71). Such an excellent outcrop condition is quite un-

6.7 Suweolbong Geosite

Fig. 6.68 Geologic and topographic map of Suwolbong at the western margin of Jeju Island

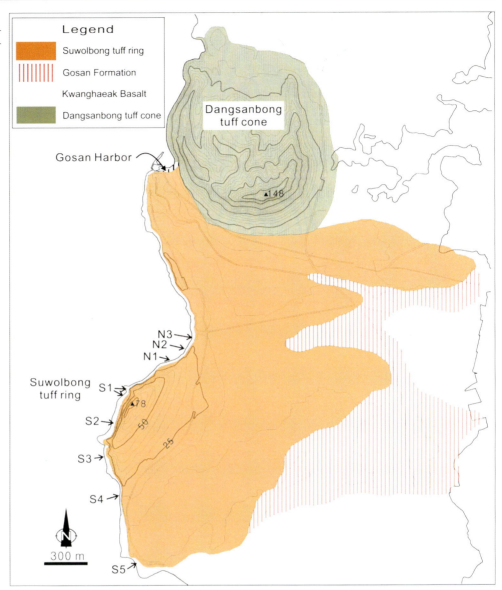

Fig. 6.69 A view of the Suwolbong tuff ring aboard a boat, consisting of well-bedded volcanic ash layers

Fig. 6.70 Continuous sea-cliff exposures of the pyroclastic deposit at Suwolbong

usual, and there are no other examples like this outside Jeju Island. Summation of lateral facies transitions within the deposits show that massive or crudely stratified lapilli tuff occurs in the most proximal part, which transforms downcurrent into either planar-stratified, undulatory-stratified or climbing megaripple-bedded (lapilli) tuff (Fig. 6.72). Further downcurrent, the tuff becomes mostly thinly and planar-stratified with subdued bedforms. Such lateral facies transitions suggest that a pyroclastic surge is highly concentrated

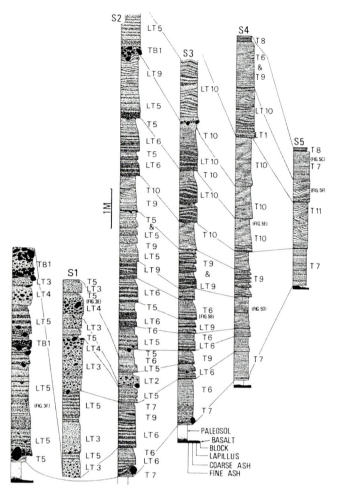

Fig. 6.71 Correlated columnar logs of the Suwolbong tuff ring, showing proximal-to-distal variations of deposit features within pyroclastic surge and related fallout deposits. (after Sohn and Chough 1989)

Fig. 6.72 Climbing megaripple bedforms within the pyroclastic surge deposits of the Suwolbong tuff ring

Fig. 6.74 Doyoji, a traditional kiln site near Shindo-ri

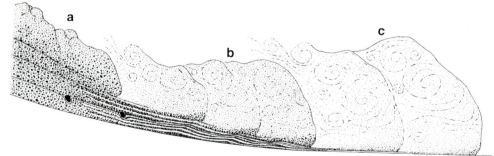

Fig. 6.73 Illustration of a pyroclastic surge at three stages (**a–c**) of development. The pyroclastic surge expands, decreases in particle concentration and develops turbulence (after Sohn and Chough 1989)

near the vent and deposits its load rapidly from suspension with meager tractional transport, resulting in the generally unstratified deposits in the proximal area (Fig. 6.73). As the surge becomes diluted downcurrent through fallout of suspended loads and mixing of ambient air, it becomes turbulent and segregated into coarse-grained bedload and overlying fine-grained suspension, forming thinly stratified units with diverse bedforms. Further downcurrent, the surge may be either cooled and deflated or pushed up into the air, depending on its temperature.

The above model for the deposition from pyroclastic surge is the first depositional model of pyroclastic surge based on sedimentological study of pyroclastic surge deposits. The model is introduced in a number of geological literatures and textbooks (Walker and James 1992; Reading 1966; Sigurdsson et al. 2000, among others. The Suweolbong tuff ring is

6.7 Suweolbong Geosite

Fig. 6.75 Gosan-ri prehistorical (the Old and New Stone Ages) site (**a**) and the excavated earthenware remains (**b**) exhibited at the Jeju National Museum

Fig. 6.76 Eongdeokdongsan (**a**) and a rock body engraved with Chinese characters denoting 'Jeolbuam' (**b**)

Fig. 6.77 Remains of Japanese military caves on the sea-cliff of the Suwelbong

Fig. 6.78 Various shapes of Dolhareubang, a traditional stone statue of Jeju Island

thus very famous in the volcanological community and is very suitable for geological field excursions.

The age of the eruption of the Suweolbong tuff ring was revealed by recent OSL (optically stimulated luminescence) dating of accidental quartz grains in the tuff (Cheong et al. 2007). The quartz OSL ages of 18.3 ± 0.7 ka and 18.6 ± 0.9 ka suggest that the eruption occurred during the last glacial maximum (LGM) when the sea level was far below the present one. The pyroclastic deposits of the Suweolbong tuff ring contain considerable amounts (>10 vol. %) of accidental components such as granule- to boulder-size, angular crystalline basalt clasts, sandstone and mudstone blocks, coarse- to fine-grained quartz sands, shell fragments and rare plutonic- and metamorphic-rock fragments (quartzite, granite, etc.) (Sohn 1992). It is inferred that the hydrovolcanic explosions at Suweolbong excavated at least 300 m or more of the volcanic, sedimentary, and plutonic/metamorphic rocks below the pre-eruption surface (Sohn 1996).

Fig. 6.79 Various types of Jeju Doldam (stone wall): *Batdam* as a field fence (**a**), *Uldam* as a fence of house (**b**), and *Sandam* as a graveyard fence (**c**)

Fig. 6.80 Jeju Haenyo, a female diver (**a**) and the Haenyo Museum in Sangdo-ri (**b**)

Fig. 6.81 Samseonghyeol, the sacred site of the myth of the three family names in the ancient ages of Tamna Kingdom

6.7.2 Cultural Heritage: Shindo-ri Doyoji (Jeju Province Monument No. 58-4)

Doyoji, also called 'Gamateo', is a kiln site for the baking of earthenware, porcelain, pottery or clayware. A kiln was needed and built in these types of particular locations, as it was easy to obtain clay and firewood, and to transport finished products. Pottery production of Jeju was very active in the areas around Daejeong-eup, Hangyeong-myeon and Aewol-eup. Located approximately 3 km east of Suweolbong, Shindo-ri Doyoji in Daejeong-eup there exists one of the best preserved traditional kilns remaining in Jeju (Figs. 6.74. Also called 'Ilgopdeureunoranggul', it is 16 m long, 1.8 m tall, and 4.7 m wide. Along with being in excellent condition, it is of significant academic value in that Shindo-ri Doyoji has typical characteristics of a stone kiln whose oven, wall, fireplace, entry, etc. are made up of stone indigenous to Jeju and is unique in style when compared to others throughout the world.

6.7.3 Archeological and Historical Heritages

6.7.3.1 Prehistoric Remains in Gosan-ri (Historical Remains No. 412)

Gosan-ri relics are located near the Jagoonae port in Gosan-ri, Hangyeong-myeon, Jeju City (Fig. 6.75). These are the only existing prehistoric remains that show the aspects of the shift from the late Old Stone Age to the early New Stone Age and the cultural characteristics of the early New Stone Age. Based on four research excavation periods, the site has been verified to possess traditional stone tools from the late Old Stone Age, and ancient earthenware, mixed with vegetable fiber, from the New Stone Age. The earthenware artifacts date back 10,000–12,000 years ago. In fact research

6.7 Suweolbong Geosite

Fig. 6.82 Samyang-dong prehistorical (the Bronze Age) site (**a**) and the excavated earthenware remains (**b**)

shows that the maximum age by radiocarbon dating is up to 10,180±65 B.P.

The culture of Gosan-ri relics represents the existence of the culture of the early New Stone Age, in which the traditional 'Jomdolnal', or a tiny stone blade from the late Old Stone Age, was used to make various tools and to produce the earthenware of Gosan-ri style that is a primitive vessel baked with low temperature for the first time in Korea. The population of Gosan-ri was regarded as the first inhabitants, establishing their community long before the agricultural society began in Jeju. They are responsible for the invention and use of the hunting arrow as applied with thrusting power over a distance as well as for use as a spear for fishing. Gosan-ri relics are currently on display at the Jeju National Museum.

6.7.3.2 Jeolbuam (Jeju Province Monument No. 9)

Located on the Yongsuri port, north of Suweolbong, Jeolbuam is known for harboring a particularly sad story of how a Ms. Goh, a virtuous lady, killed herself after waiting for her husband who had gone missing while fishing out at sea. Ms. Goh, born in Chagui-chon at the end of Joseon Dynasty, had married to a fisherman named Gang Sa-chul at the age of 19 and settled down to make a home. One day her husband went missing and Ms. Goh, in extreme grief, gave up eating and drinking and wandered along the coast for many days and nights in search of his body. With no success, after ten days and in despair, she hung herself from a hackberry tree on the cliff known as 'Eongdeokdongsan' near the Yongsuri beach. Oddly enough the next day, the body of her husband rose out of the water at a point just under the tree. Hearing this story, Judge Shin Jae-woo had 'Jeolbuam' carved on a stone, allowing the Yongsuri people to have a symbol for appeasing their sorrow. To this day there are ceremonies being held in Yongsuri on the 15th of May of the lunar calendar (Fig. 6.76).

6.7.3.3 Japanese Military Advance Base at Suweolbong

On the shores of Hanjang-dong, Gosan-ri, is a former Japanese military site considered an "an advanced base" designed to defend against any U.S. attack approaching from the western side of Jeju toward the end of the Pacific War. According to a witness from the adjacent neighborhood the tunnel in

Fig. 6.83 Jejumok government offices, the hub of politics, administration and culture of Jeju Island during the Joseon Dynasty

Hanjangal (Fig. 6.77) was dug by a Japanese expeditionary platoon from Gama Oreum. After the completion, the Japanese soldiers set up artillery and often took up target practice at objects floating in the sea. Lighting equipment was place within the dozens of holes that were drilled along both sides of the tunnel. There are vivid traces of pickaxing and iron materials similar to rusty nails embedded in the wall. A bunker made up of gravel and cement was built toward the end of the tunnel. The tunnel reaching the tochka has four different spaces. The tochka is 3 m wide and has a hole 70 cm wide and 60 cm high at the front. There are pieces of plywood attached to the remaining wall.

6.7.3.4 Japanese Military Suicide Squad Position at Suweolbong

The point at low-tide along the shore of Suweolbong reveals a taxiway designed for Japanese Army suicide-boats built during the end of the Pacific War. The taxiway was made up of gravel and cement. At one time it had been connected with the suicide-boat shed, or the tunnel. Based on the fact that there is a 60 m distance from the taxiway remains to the tunnel on the shore, it seems that the taxiway initially had a minimum length of 70 m.

6.7.4 Legend of Suweolbong (Nokgomoru)

Suweolbong is a high hill located in the western tip of Jeju Island, so called 'Nokgomoru'. About 350 years ago, there was a girl named Suweol and her brother Nokgo, who lived there taking care of their mother. Although they lost their father at an early age, and were poor, Suweol and Nokgo made a good home. One spring, their mother fell ill and, despite all medicines, she became worse every day. Suweol and Nokgo exhausted all attempts and could do nothing but cry. One day a Buddhist monk was passing by and, learning of their plight, he suggested to boiling 100 herbs and then giving it to their mother. From that day, Suweol and Nokgo started collecting the herbs but they couldn't find a final one known as Ogalpi'. According to the monk, this herb grows atop rocks or on a steep mountain slope. The next day, Suweol and Nokgo needled their way through the hill along the seashore. Looking carefully, they finally found the herb. 'Ogalpi' was stuck on the middle of the cliff. Without fear, and with enthusiasm, they climbed down the cliff in pursuit. Suweol stepped down holding Nokgo's hand and picked the herb. When she handed it over to Nokgo, Suweol was too excited and lost her balance. Nokgo, too, full of joy, couldn't hold his grip on her hand And Suweol fell down the side of the steep cliff. Finally realized what had happened, Nokgo screamed, almost losing his mind, and couldn't stop crying out. His tears ran through between rocks and gushed out.

Over the ages, the hill was called 'Nokgomul Oreum', 'Mulnari Oreum', 'Nokgomoru', or 'Suweolbong'.

6.8 Other Cutural Heritages

6.8.1 Dolhareubang

Dolhareubang is a stone statue carved out of Jeju basalt (Fig. 6.78). It is the most commonly identifiable representative artifact and folk-craft item known to Jeju. Dlehareubang were originally placed at the entrance to castles of Jeju-mok, Jeongui-hyeon and Daejeong-hyeon, playing the role of guardian. The one of Jeju-mok was called as 'Wooseokmok', one of Jeongui-hyeon as 'Beoksoomeori' and one of Daejeong-hyeon as 'Mooseonmok'. Occasionally it was called 'Ongjoongseok', 'Dolhareubang'. Labeled as such by children, it is typically accepted as the officially recognized name. The shape of Dolhareubang somewhat varies on regions, but it is commonly from 1.1–2.4 m high, adorned with a round helmet-style hat with a narrow forehead, bulging eyes, a bottle-shaped nose, clenched lips and, in order to show its dignity, there are two hands on the upper and lower part of the belly (Fig. 6.79).

6.8.2 Doldam (Stone Walls)

As an essential element of its culture, Jeju Doldam makes up the majority of the scenery throughout Jeju. Porous basalt is found in great abundance in this volcanic island of Jeju. Doldam is a stone wall made of Jeju basalt and has multiple uses. Doldam is categorized according to its shape and purpose. In fact, there is a great variety of Doldam: *Chugdam* as an outer wall of house; *Uldam* as a fence of house; *Olledam* as an entrance alleyway toward house; *Batdam* as a field fence; *Sandam* as a graveyard fence; *Jatdam* built in a horse farm; *Seongdam* as a protective wall at the seashore or fortress; *Wondam* as a fish-trap fence of traditional fishery. The origin of Doldam is not exactly clear, but it is assumed to have appeared following the beginning of the period of settlement and agriculture. The origin of Batdam appears, appears in an historical record, explaining that a judge Kim Goo at the time of Goryeo Dynasty let people build stone walls on the borders of farmlands to avoid the despotism of the powerful. Although it is an important function to mark the borders of farmlands, the primary role of the doldam is protection against wind. In addition to this, doldam played a role of protecting farmlands and graveyards from horses and cows grazing in the wilds of the mid-mountain area of Jeju. These walls also acted as Seongdam, functioning as defense against any foreign invaders.

6.8.4 Haenyo

Haenyo is a professional female diver who harvests marine products by going under the sea without the use of any particular breathing apparatus. Jeju Haenyo is the symbol of Jeju women's persistent vitality (Fig. 6.80) and the strong pioneer spirit in the face of the challenging and barren nature of Jeju. Originally Jeju Women Divers were called Jamnyo or Jamsu: in 'Record of Jeju Climate' written by Yi Geon in the early 17th century, Jamnyo is a woman diver who gathers seaweeds under the sea from February through May and works naked using a sickle. At the age of 7 or 8, Haenyo usually starts learning to dive shallow water and when the age of 13 through 14, she is ready to practice a professional diving. The group of Haenyo is divided to three levels of hierarchy depending on individual skill. Their sense of community is more loyal that of any other. Haenyo, from the top level of its hierarchy, can reach depths of up to 20 m for 2 min at a time.

Based on their unusual perseverance, and the power of unity, Jeju Haenyo has expanded the scope of their national consciousness through underwater harvesting, the maintenance of a fishery, the usage of professional tools and the management of the community. They have also played a leading role in creating and transmitting their own unique culture such as its special language, strong shamanism, Haenyo songs as Jeju's typical labor music and the customs specific to its community. Historically Haenyo led the anti-Japanese movement under the time of Japanese rule. During the 1930s, great numbers of Japanese fishermen arrived, over-fishing the fertile seas of Jeju. The Haenyo catch depleted and forced them to look elsewhere for more fertile fishing areas. At that time Jeju Haenyo's territories spread to various parts of Japan, Daeryeon and Chungdo in China, and even Vladivostok of Russia and beyond to include all areas of the Korean Peninsula. Today, however, following the rise of industrialization and the exhaustion of fishery resources, the number of Haenyo has decreased reaching only about 5,000. The existing group of Haenyo is quickly aging with less than 15 % of the divers being 40 years of age. Haenyo Museum in Sangdo-ri, Gujwa-eup, Jeju, exhibits remnants and documents related the history of Jeju Haenyo.

6.8.5 Bangsatap (Stone Towers) (Jeju Province Folklore Material No. 8)

These stone towers were constructed at sites where the flow of vital energy (*ki* in Korean) is stunted or inharmonious according to geomancy. Seventeen towers are recognised as significant and accorded Provincial Folklore status.

The wisdom of Jeju people consists in the culture of Doldam: at first glance, it appears loosely stacked, but in truth it is carefully placed with gaps that allow the infamous strong winds to blow through without knocking it down. Olledam as a entrance alleyway is built with a natural curve so that it can protect a house from a head wind and shield the house from outsiders' view. Not to mention, the scenery created by the various shapes of Doldam generates such an exclusive aesthetic value, it is a statement to the practical enduring beauty of the Jeju people.

6.8.3 Jeju Chilmeoridang Yeongdeunggut (UNESCO Important Intangible Cultural Heritage No. 71)

The Jeju Yeongdeunggut is a ritual held in the second lunar month, called "Yeongdeung". In honor of the goddess of wind, Grandmother Yeongdeung, it involves prayer intended to bring an abundant harvest and a plentiful catch at sea. On the first day of the Yeongdeung month, Grandmother Yeongdeung arrives with her family, the winds, to enjoy the beauty of Jeju. Spring arrives when she and the winds depart. Grandmother Yeongdeung sows the seeds of five grains and plants seaweed seeds along the shore to ensure an abundance of crops, shells, abalones and seaweed. During her fifteen-day stay, each village performs a shaman rite called Yeongdeunggut where the goddess visits each neighborhood. Of all these rites, the one called the Jeju Chilmeoridang Yeongdeunggut held at the Chilmeoridang Shrine in Sanji harbor of Jeju City, is the largest and most representative: it begins with the Yeongdeung welcoming rite on the 1st day of the second lunar month and ends with the Yeongdeung farewell rite on the 14th day.

The Jeju Yeongdeunggut is funded by either the village organizations or by the Haenyo communities and a rite officiator is chosen among the fishermen or Haenyo. It begins with Chogamje, a "calling of the gods" ceremonywhich involves inviting the gods to the shrine and wishing for the well-being of the participants, and is followed by Yongwangmaji, a "welcoming the Dragon King" to appease the gods and to send them back. Moreover, instead of the sowing of seaweed seeds, millet seeds are thrown to the sea and on the mat in order to tell Haenyo's fortune. Jidrim is then performed by tossing offerings wrapped with white paper to the Dragon King and the spirit of fishermen lost at sea. The Jeju Chilmeoridang Yeongdeunggut was officially listed as UNESCO's Intangible Cultural Heritage of Humanity in September 2009.

6.8.6 Samseonghyeol (Historical Remains No. 134)

This site, with its distinct depression and traces of three holes, demonstrates the ongoing link between geology and culture so vividly evident on Jeju. Samseonghyeol, located in Ido-dong, Jeju City (Fig. 6.81), is the mythical birthplace of the Jeju people and their culture. It is the sacred site of the myth of the three family names which tells a story of three demi-gods, who came from within the earth and founded the Tamna Kingdom. There are still three holes remaining, which are believed to be the birthplace of these three demi-gods. The people of Jeju have passed the myth of the three family names down over the centuries by word of mouth. It is also found from a couple of ancient documents, such as Goryeosa, Yeongjuji, etc.

The myth of the three family names is as follows: in the beginning man did not exist in Jeju. All of sudden, one day, three demi-gods named Eulla rose from Samseonghyeol and became the originators of three family names, Goh, Yang and Boo. They lived in harmony, sharing hunting within their carnivorous culture. One day a box was discovered drifting down from the East Sea. On opening the box they discovered three princesses from Byeokrang (a fictional state) in the East Sea with a calf, a colt, and the seeds of five individual grains. Each of the demi-gods married one of the princesses; according to each one's age. They each shot an arrow, settled wherever it landed, and began planting crops and raising animals. They soon prospered in their new agricultural economy and the island developed into the Tamna Kingdom.

The myth not only expresses the origins of these family names but it also represents the birth of the nation, including the founding of a political system and its rulers. Although the site of Samseonghyeol had been abandoned for a long time, in 1526, to honor the memory of its founders, Jeju Governor Yi Soodong erected monuments and had protective stone walls around the city, passable though a large red gate, and allowed descendents to participate in sacrificial ceremonies. Since then the site of Samseonghyeol has been furnished with Samseongjeon, Jonsacheong, Sungbodang and its affiliated facilities. It is surrounded by lush forests inhabited by a dozen varieties of trees which are more than 200 years old.

6.8.7 Prehistoric Remains in Samyang-dong (Historical site No. 416)

Located in Samyang-dong, Jeju City, this prehistoric site contains the earliest known and largest residential remains built in Jeju, dating back to the 3rd century B. C. (Fig. 6.82). The ancient village in Samyang-dong covers an area of approximately 10 m²— a coastal plain area occupied with both large and small dugout huts, a storehouse, a warehouse, a kiln for earthenware, a kitchen as well as a stone embankment dividing each space in the village, a drainage system, a refuse dump, a shell mound, a dolmen and so on. Excavation research has confirmed a total number of 236 dwellings. This research has discovered various earthenware and tools such as a stone sword, a stone arrow, a bronze sword, a bronze arrow, a jade bracelet, ironware, a glass bead, a stone ax, a stone plane and a whetstone, burnt crops like rice, barley, bean, a nutmeg nut, an acorn, a peach, and seeds, etc. There are small squares for outdoor gathering in the middle of the site. There is also a housing area arranged as a large-sized dwelling site with about 6 m in a unit, as well as 12–15 small-sized residential sites. The essential relics such as a variety of adornments, a bronze sword and a large-sized vessel are only excavated in the large-sized residential site. Therefore it is believed that the layout of residential areas and the type of relics reflects a structure of a hierarchical society of that time, where layout varied according to a social position. The site of the residential areas, and a part of the relic, are currently displayed in the Samyang-dong Prehistoric Historical Site and visitors can experience the life of the time through 14 dugout huts restored to their original condition.

6.8.8 Jejumok Government Office Buildings (Historical Remains No. 380)

The Jejumok government offices are located inside of Jeju castle (Fig. 6.83). The site had been the hub of the history and administration of Jeju since the Tamna State was established. The historical district is based on the old Daechon-hyeon, which were housed with the Jejumok offices during the Joseon Dynasty and is known as the site of the government offices of the Tamna State. This area, chosen as the capital or central area, grew prosperous while Jeju was partitioned into Jejumok, Daejeong-hyeon and Jeongeu-hyeon in 1416. The Jejumok offices were centralized on the left and right side of Gwandeokjeong (Treasure No. 322), which is the southern and the northern area of the city. In particular, the minister's office and its annex buildings (the restored area) were mostly in north along with judicial offices. The main facilities include Yeongbingwan (guesthouse), Honghwagak (main office), Yeonheegak (minister's office), Jongru (gate of a minister's office), Mangkyeonru, Aemaeheon, Gullimdang and Chanjuheon (judge's office). It is confirmed that the unit of government office buildings had been renovated three times. According to the record, Jejumok Gwana of the early Joseon Dynasty had a total of 58 units and 206 rooms.

Future Geosites

7

There are many sites which deserve to be identified, managed and interpreted as geosites within the Jeju Island Geopark. Twenty-one sites have been identified for inclusion over the next 10–15 years (Fig. 7.1). An Action Plan will be developed to progressively formally incorporate these additional sites within the Jeju Island Geopark. Other geosites may be added from time to time as research demonstrates their values and the practicality of opening them to the public is assessed.

7.1 Dangsanbong Tuff Cone

Dangsanbong is a tuff cone situated on the western margin of Jeju Island. It has a horseshoe-shape morphology with an opening toward the north and a nested scoria cone at its center (Fig. 7.2). The crater rim is more than 900 m wide and as high as 148 m above present sea level. A geological study (Sohn and Park 2005) suggests that the tuff cone is one of the oldest volcanic formations in Jeju Island, overlain by later lava flows. The tuff cone is mainly composed of steeply inclined strata of lapilli tuff that are inward and outward dipping. The tuff cone strata are divided into two distinct stratal packages by a volcano-wide truncation surface (Fig. 7.3a). The lower stratal package (LSP) beneath the truncation surface consists almost entirely of steeply inclined and outward-dipping beds that dip generally between 20 and 30°. On the other hand, the upper stratal package (USP) consists of inward-dipping beds with their outward-dipping counterparts mostly removed by erosion. Contrasting lithofacies characteristics between these stratal packages (Fig. 7.3b, c) are interpreted to have resulted from a change in eruption style of Dangsanbong from a cone-forming to a ring-forming (surge-dominated to be more exact) eruption, probably associated with a volcano-wide collapse event (Sohn and Park 2005).

A wedge-like sequence of very poorly sorted, disorganized to very crudely stratified bouldery deposits named the Gosan Formation accumulated around the Dangsanbong tuff cone (Fig. 7.4). Overall reddish or brownish coloration of the formation as well as the overall structures and textures of the deposits suggests that the formation is the deposit of an ancient scree developed around the dissected Dangsanbong tuff cone. The quartz OSL age of the formation is 23.2 ± 1.0 ka (Cheong et al. 2007).

7.2 Chagui Island

Chaguido is a small island, about 0.16 km^2 in area, located off the western coast of Jeju Island (Fig. 7.5). The island comprises three main islets and dozens of small rocks. Although the island is not inhabited, it is renowned as an excellent fishing and diving point on Jeju Island. Although geological studies have not yet been carried out, the island is regarded as a promising geosite in the future because it provides excellent outcrops of all types of basaltic rocks, including steeply inclined-bedded lapilli tuff, reddish scoria deposits, agglutinates, lava flows, dikes, and reworked volcaniclastic sedimentary rocks (Fig. 7.6), suggesting complex monogenetic volcanism involving both magmatic and phreatomagmatic eruptions.

7.3 Biyang Island

Biyangdo is a small island, located off the northwestern coast of Jeju Island (Fig. 7.7). The island comprises a scoria cone with well-preserved morphology. The island is famous as a possible site of a historic eruption about one thousand years ago, although the assumption needs to be verified by further geological studies. The island is composed mainly of scoria deposits and agglutinates, which resulted from a Hawaiian eruption (Fig. 7.8). A strange rock formation, named Aegieopeundol with a meaning of a woman carrying a baby on the back, is found along the shore (Fig. 7.9). The rock formation, designated as a natural monument, is interpreted to be a hornito, which is a small opening or rootless vent that releases small quantities of lava when high pressure within a lava flow causes lava to ooze and spatter out.

K. S. Woo et al., *Jeju Island Geopark—A Volcanic Wonder of Korea,* Geoparks of the World,
DOI 10.1007/978-3-642-20564-4_7, © Springer Verlag Berlin Heidelberg 2013

Fig. 7.1 Proposed geosties within the Jeju Island Geopark in the future

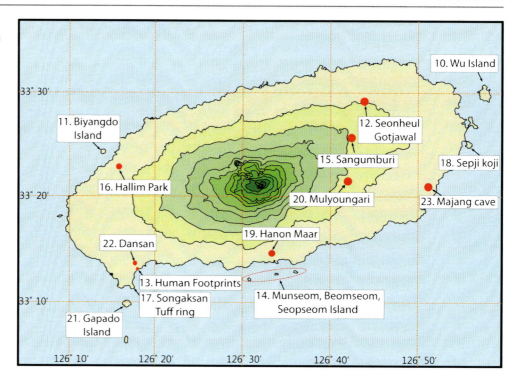

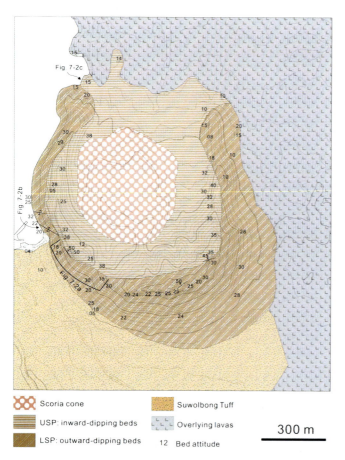

Fig. 7.2 Geological map of the Dangsanbong tuff cone with a nested scoria cone at its center

7.4 Sangumburi Crater

Sangumburi is a peculiar volcanic crater, which is distinguished from the other volcanic craters and edifices on Jeju Island. The crater rim is 31 m higher than the surroundings; the rim-to-rim width is 635 m; the diameter of the crater floor is about 300 m; the height from the crater floor to the rim is 132 m (Fig. 7.10). The crater has been known among Koreans as a 'maar' based on its morphology for many decades. The information boards within the Sangumburi Park as well as a number of books and websites also introduce the crater as the only one 'maar' on Jeju Island. Recent study shows, however, the area around the crater is composed only of lavas (Fig. 7.11). Lavas together with some clinker are exposed along the trails around the crater and on the inner wall of the crater, suggesting that there were only effusion of 'aa' lavas. No ejecta beds have been found, negating the possibility of explosive excavation of the crater by 'maar'-forming processes. The Sangumburi is therefore interpreted to be a pit crater formed by sinking or collapse of the surface surrounding a vent for lava.

7.5 Geomunoreum Scoria Cone

Geomunoreum is a fairly large scoria cone in the northeastern part of Jeju Island. The scoria cone has a horse-shoe shape due to break-through of the lava flows toward the northeast direction (Fig. 7.12). Lava flows from the scoria cone are believed to have flowed down the slope of Mt. Hallasan

7.5 Geomunoreum Scoria Cone

Fig. 7.3 Outcrop features of the Dangsanbong tuff cone. **a** A volcano-wide truncation surface between the upper and lower stratal packages (*USP* & *LSP*). **b** Thinly stratified lapilli tuff of LSP with a V-shaped chute at center, suggesting emplacement by grain flows with a short period of erosion. **c** Cross-stratified tuff of *USP*, suggesting emplacement by powerful pyroclastic surges

Fig. 7.4 A wedge-like sequence of scree deposits (*Gosan Formation*) derived from the Dangsanbong tuff cone and then later overlain by the tuff from the Suwolbong tuff ring

Fig. 7.5 A view of Chagui Island composed of three main islets and dozens of small rocks

Fig. 7.6 A coastal exposure of volcanic rocks at Chagui Island composed of steeply inclined-bedded lapilli tuff, reddish scoria deposits, agglutinates, lava flows, and dikes

in a north-northeast direction down to the coastline for about 13 km, forming a series of lava tubes, including Seonheul Vertical Cave, Bengdwigul Lava Tube, Bukoreumdonggul Lava Tube, Daerimdonggul Lava Tube, Mangjanggul Lava Tube, Gimnyeonggul Lava Tube, Yongcheondonggul Lava Tube, and Dangcheomuldonggul Lava Tube towards the sea (Hwang et al. 2005), which are mostly designated as national monuments. Because of the geological significance, the scoria cone was also designated as a natural monument in 2005. Pine trees, Japanese cedars, and Oriental arbor vitaes are growing over the scoria cone, making a dense forest.

There are also a number of historical sites within the scoria cone, which originated from the Japanese colonial period (Fig. 7.13).

7.6 Dusanbong Tuff Cone

Dusanbong is a relatively old tuff cone in the eastern part of Jeju Island (Fig. 7.14). It comprises a scoria cone within its crater and has an opening toward the west, which acted as an outlet of a lava flow. The outer rim beds of the tuff cone were mostly removed by erosion probably during the last interglacial. Nevertheless, the diameter of the volcanic edifice exceeds 1.2 km, suggesting that it was originally a relatively large tuff cone. The overall stratigraphy of the volcano is almost identical to that of the Songaksan tuff ring and the Udo tuff cone, suggesting an eruption that changed in eruption style from phreatomagmatic to magmatic. Dusanbong recently became a famous site because it is the starting point of the very popular Olle tracking courses on Jeju Island and because of the breathtaking sceneries from the top of the tuff cone (Fig. 7.15).

7.7 Udo Tuff Cone (Someorioreum)

Udo is a small island, about 3×4 km across, located ~3 km east of Jeju Island (Fig. 7.16). It comprises a tuff cone, a younger nested spatter cone, and overlying basaltic lava

Fig. 7.7 An aerial view of Biyang Island composed of a scoria cone

7.7 Udo Tuff Cone (Someorioreum)

Fig. 7.8 An outcrop of crudely bedded agglutinate on Biyang Island

Fig. 7.10 An aerial view of the Sangumburi Crater

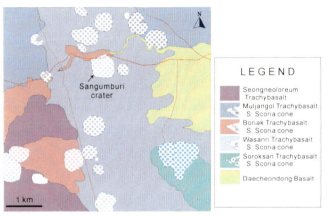

Fig. 7.11 A geological map of the area surrounding the Sangumburi Crater

Fig. 7.9 A strange rock formation, named Aegieopeundol, which is interpreted to be a hornito

Fig. 7.12 Digital elevation model of the Geomunoreum scoria cone

shield. Dating of the lava shield rocks gave a K–Ar age of 114 ± 3 ka, whereas dates of core samples gave $^{40}Ar/^{39}Ar$ ages of 102 ± 69 and 86 ± 10 ka (Koh et al. 2005, 2008). The tuff cone, named the Someorioreum, is horseshoe-shaped with a rim-to-rim width of 800–900 m and a height of 132 m. The tuff cone generally comprises steeply inclined (20–30°) beds of lapilli tuff and tuff that dip radially away from the vent (Fig. 7.17). A detailed sedimentological study of the tuff cone reveals that it has formed by a Surtseyan-type eruption, which became drier towards the end of the eruption (Sohn

Fig. 7.13 A Japanese encampment within the Geomunoreum scoria cone, which was built during the Japanese colonial era

and Chough 1993). The deposition was mostly accomplished by grain flows of lapilli and blocks in addition to airfall of finer-grained tephra. Absence of marine-reworked deposits suggests that the majority of the tuff cone was constructed subaerially, although the submerged part may have formed underwater. Common inclusion of acidic volcanic rock fragments (rhyolite and welded tuff) that were most likely derived from the Cretaceous volcanic basement rocks in the eastern Jeju area suggests that the level of hydrovolcanic explosions and the depth of country rock excavation reached more than 300 m below the present sea level (Sohn 1996). Recent high-resolution geochemical study (Brenna et al. 2010) suggests that the eruption of Udo began with relatively evolved alkali magma. The magma became more primitive over the course of the eruption of the tuff cone, but the last magma to be explosively erupted had shifted back to a relatively evolved composition. A separate sub-alkali magma batch was subsequently effusively erupted to form a lava shield.

Fig. 7.14 A view of the Dusanbong tuff cone from the south

Fig. 7.15 A view of the eastern part of Jeju Island from the top of the Dusanbong tuff cone, characterized by numerous scoria cones

Fig. 7.16 An aerial view of Udo Island from the south

Fig. 7.17 An excellent coastal exposure of the tuff cone deposits overlain by basaltic lava flows and reworked volcaniclastic deposits at Udo Island

7.8 Oedolgae

Oedolgae is a 20 m high sea stack near the south-central coast of Jeju Island (Fig. 7.18). It is composed of trachytic lava flow that is hundreds of thousand years old. The sea stack has a legend of a woman who turned into a stone statue after waiting for her fisherman husband for a long time. The Oedolgae coast also has excellent geological exposures of peculiar volcanic rocks that are related to intrusion of magma, mixing of magma and unconsolidated sediments, and escape of heated steam and gas from the wet sediment (Fig. 7.19).

Fig. 7.18 A view of the Oedolgae sea stack at the south-central coast of Jeju Island

Fig. 7.19 An exposure of peculiar volcanic rocks near the Oedolgae sea stack that formed in association with intrusion of magma into wet and unconsolidated sediments

7.9 Songaksan Tuff Ring

The Songaksan is a tuff ring located at the southwestern margin of Jeju Island (Fig. 7.20). It comprises a scoria cone and a ponded lava flow in its crater, (Figs. 7.21, 7.22). The rim beds of the tuff ring are up to 80 m thick and extend northward and northwestward for more than 2 km. The rim-to-rim width is estimated to be about 7–800 m. The tuff ring comprises mainly thin-bedded and gently dipping tuff with abundant megaripple bedforms emplaced by pyroclastic surges (Chough and Sohn 1990). The deposits are composed of mainly sideromelane/tachylite ash and some poorly vesicular lapilli that have blocky equant shapes, as well as considerable amounts of accidental components, which suggest deep excavation (more than ~300 m) and incorporation of abundant country rocks by hydroexplosions. The lithofaci-

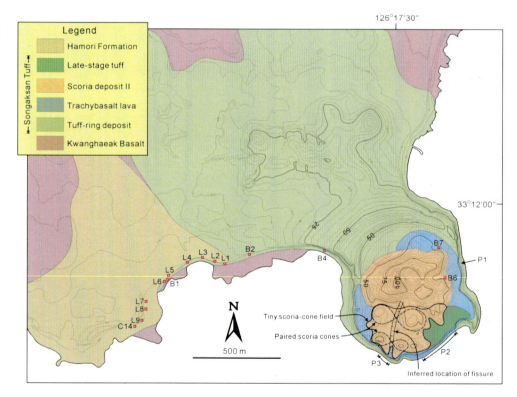

Fig. 7.20 A geological map of the Songaksan tuff ring located at the southwestern margin of Jeju Island

Fig. 7.21 A distant view of the Songaksan tuff ring from the west, consisting of a tuff ring, a nested scoria cone and a ponded lava flow

7.10 Dansan Tuff Ring/Cone

Fig. 7.22 A seacliff exposure of the volcanic rocks at the Songaksan tuff ring. *TR* = tuff ring, *SC I* = scoria deposit I, *TBL* = trachybasalt lava, *SC II* = scoria deposit II

es characteristics of the Songaksan tuff ring are generally similar to those of the Suwolbong tuff ring. There are, however, several subtle differences in facies characteristics, which suggest that the pyroclastic surges at the Songaksan tuff ring were relatively wetter and less energetic than those of the Suwolbong tuff ring (Sohn 1996).

7.10 Dansan Tuff Ring/Cone

Dansan is an old volcanic edifice located in the southwestern part of Jeju Island. It consists of two differently oriented ridges that have different lithofacies characteristics and bed attitudes (Fig. 7.23). The northern ridge is sharp-crested, arcuate in plan and rises more than 100 m above the surrounding lavas whereas the southwestern ridge has a relatively low relief, protruding about 40 m above the surrounding lavas. The overall characteristics of Dansan, composed of two different rim beds indicating different source vent directions,

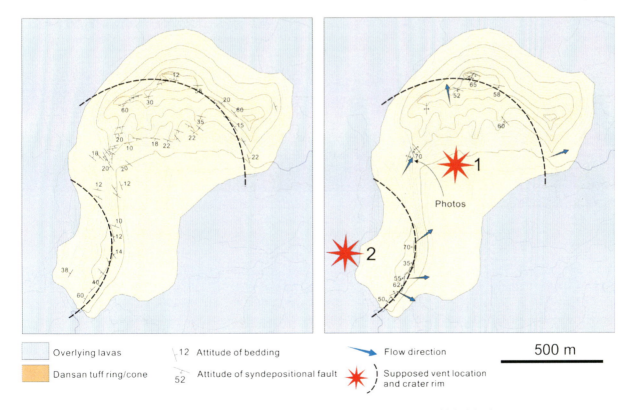

Fig. 7.23 A geological map of the Dansan tuff ring/cone complex at the southwestern part of Jeju Island

Fig. 7.24 An outcrop showing the contact relationship between two rim deposits at Dansan, which originated from different source vents

Fig. 7.25 A view of the Seopjikoji promontory, consisting of a dissected scoria cone and overlying lava flows

Fig. 7.26 An outcrop of the Seopjikoji coast, showing agglutinated scoria/spatter deposits and a small dike

suggest that the volcano originally comprised a pair of juxtaposed craters probably with a figure eight configuration in plan. The contact relationship between the two rim deposits (Fig. 7.24) shows that they formed sequentially (from vent 1 followed by vent 2) in response to migration of the active vent with an intervening erosional break. The contrasting lithofacies characteristics between the two rim deposits suggest that there was a significant change in the eruption style associated with the vent migration from a cone-forming (fallout-dominated) to a ring-forming (surge-dominated) eruption (Sohn and Park 2005).

7.11 Seopjikoji

Seopjikoji is a small promontory at the eastern margin of Jeju Island and to the south of the Ilchulbong tuff cone (Fig. 7.25). This area was one of recent volcanic centers on Jeju Island, comprising a scoria cone and several lava flows. These volcanic formations were dissected by marine waves, exposing excellent outcrops of bedded scoria deposits, agglutinates, dikes, and a variety of lava flows (Fig. 7.26) and generating lovely coastal landscapes. This area has therefore been used as the shooting site of a number of films and TV dramas and attracts millions of visitors every year.

Fig. 7.27 A view of the Hanon Crater

Geotourism

8

Education opportunities with and about geoparks are to provide and organize support, tools and activities to communicate scientific knowledge and environmental concepts to the visitors (e.g. through museums, interpretive and educational centers, trails, guided tours, popular literature and maps, modern communication media and so on). They also allow and foster scientific research and cooperation with universities, and between geoscientists and local people. All educational activities should reflect the ethical considerations around holistic environmental protection and sustainable development. One of the main issues is to link education in a local context with all stakeholders. As an example of the importance of Jeju Island as an educational destination, one third of all Korean school excursion students visit Jeju (about 603,000 students in 2011).

The success of geopark educational activities depends on: the content of programs, competent, knowledgeable staff to interpret the geosites and their significance, efficient logistic support and access for the visitors, effective experiential activities, and personal interactions with the local population, experts and decision-makers.

Desired outcomes include: better geological knowledge and awareness, environmental consciousness and understanding the cultural setting of geosites, elevated visitors' satisfaction and interests, better conserved geoheritage while at the same reinforcing local awareness, pride and self-identity, and support for the development of sustainable regional economies.

Methods of conveying the geoheritage knowledge and awareness include: (1) brochures, other written material, multi-media presentations and so on, (2) information panels, (3) models, (4) hands-on activities, (5) indoor classes, (6) excursions, (7) seminars and workshops, and (8) scientific lectures.

A number of organizations have been heavily involved in facilitating environmental education for local groups and for those visiting from other areas outside Jeju Island. These organizations have provided a wealth of environmental education activities which have taken place in the area for the following groups such as nurseries and preschools, primary and secondary schools, youth groups, colleges and universities, adult education groups, retired groups, professional organizations, tour groups, and local residents.

These will continue to be the target audiences but with a special emphasis on improving the educational outcomes for school groups and by providing better, more sophisticated geosite interpretation for domestic and international tourists.

8.1 Education Facilities

There are a very large number of existing facilities on the island related to nature-based tourism as well as plans for others. The Jeju Special Self-Governing Province, through institutions such as the Jeju National Museum and the World Heritage management office, has a comprehensive program of training for heritage interpreters and guides.

These facilities include Folklore and Natural History Museum with 30 year tradition, Jeju Stone Culture Park combining Jeju stone culture and natural history and a few visitor centers located in geosites. World Heritage Center was established in 2012 near the Geomunoreum (the source of the lava flows that formed several lava tube caves) which is a part of Natural Heritage in the World Heritage List. This center includes scientific displays and carries various functions such as education, promotion and scientific activity.

8.1.1 Visitor Centers and Visitor Points

A number of contact points exist or are planned for the Jeju Island Geopark as shown in Fig. 8.1. Four Geopark Visitor Centers will be supplemented by Visitor Points. The Geopark Visitor Centers will be established to provide better service to visitors and to protect and make use of geosites effectively. They are to provide information, such as geological knowledge, trail maps, camp sites, staff contact, lodging facilities, restaurants and other items relevant to tourism. They will offer in-depth educational exhibits and artifact displays on, for example, geological history, volcanic activity, land-

K. S. Woo et al., *Jeju Island Geopark—A Volcanic Wonder of Korea*, Geoparks of the World,
DOI 10.1007/978-3-642-20564-4_8, © Springer Verlag Berlin Heidelberg 2013

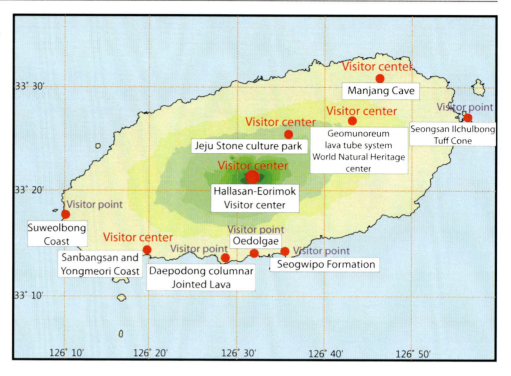

Fig. 8.1 Existing and proposed visitor centers and visitor points in Jeju Island Geopark

scape development and natural history. Often film or other media display is used. If the site has permit requirements or guided tours, the visitor center is often the place where these are coordinated.

In addition to these contact points at the geosites, two Jeju Island Geopark information centers will be established in Jeju and Seogwipo cities. This will allow inbound visitors and city residents to obtain information about the geopark and the individual geosites and allow them to plan their visits effectively.

The Global Geoparks Network suggests that a geopark should achieve its goals through a three-pronged approach of Conservation, Education and Tourism. Visitor centers are an essential facility supporting this approach. The roles and functions of the geopark visitor centers will be: (1) to provide basic information by informing visitors and the local community about the geopark and its geosites and by giving objective information on geology, history, culture, and scientific knowledge generally, (2) to provide educational and experiences using exhibits and artifact displays with multi-media, educational and experience program focused on target group, and guided programs, (3) to protect, supervise and monitor the geosites by taking the initiative to protect and supervising the geosites, by monitoring geological and other environmental changes, and by raising environmental awareness, and (4) enhance the network with local community by providing networking programs with the local tourism industry, by providing educational programs to local residents by guides and volunteer georangers, by displaying and sel-

ling appropriate local products and crafts, and by creating and revitalizing socio-economic opportunities.

In Fig. 8.1 four visitor centers will be supplemented by visitor points. The visitor centers will be multi-function visitor center in which the distinctive features of geosites are emphasized. They will provide information on the Jeju Island Geopark, specific information on all nine geosites but with emphasis on the geosites in their locality, exhibition, educational and experience programs, guided tours, protection, supervision and monitoring of the geosites, and display and sell appropriate local products and crafts.

The five Visitor Points will provide more specific and will concentrate on the local geosites. They will provide simple exhibitions, guided tours, and display and sell appropriate local products.

The various features and functions of the visitor centers and points respectively are set out in Table 8.1. They are intended to implement a qualitative improvement of services to the visitors. They will provide not just simple consumptive geotourism but will emphasize discovering the importance and values of geosites. They will try to change geosites from passive tour places to active geoheritage experiences. Visitor centers will be well-organized and will enhance the systematic administration and management of the geosites. Hopefully, they will raise geological and environmental consciousness through the educational activities and publicities through investing geosites with new value and meaning and elevating visitor satisfaction by active management to suit various interests and needs. Visitor centers may be able to strengthen connections with local community by involving local people as a most important factor for the successful

8.1 Education Facilities

Table 8.1 Features and functions of geopark visitor centers and visitor points

Feature	Function	Visitor center	Visitor point
Exhibition hall	Displays	Yes	Yes
Multi-media room	Information and exhibitions using multi-media	Yes	Yes
Conference room/ auditorium	Educational programs	Yes	No
	Guide and volunteer programs	Yes	No
	Experiential programs	Yes	No
	Seminars and special lectures	Yes	No
Learning room(s)	Guided tours	Yes	Yes
	Guide and volunteer facilities	Yes	Yes
	Educational programs	Yes	Yes
Lobby/shop	Special events	Yes	No
	Geopark-appropriate souvenirs and local products	Yes	Yes

Fig. 8.2 Eorimok visitor center at Mt. Hallasan

Fig. 8.3 Geological display at the Eorimok visitor center

Fig. 8.4 Biological display at the Eorimok visitor center

establishment and maintenance of a Geopark and by revitalizing local economy through tourism based on local resources.

Some of the geosites, other than the World Heritage sites, have small tourist information kiosks on site. The role of these will be expanded to enhance the geopark concept. The three World Heritage sites, the Hallasan Geosite Cluster, Manjang Cave and the Seongsan Ilchulbong Tuff Cone have guided tours, more comprehensive information and enhanced visitor facilities. Those at the Hallasan National Park headquarters at Eorimok are particularly well developed and include a new visitor center to interpret the World Heritage values of Jeju Island particularly of Hallasan.

8.1.1.1 Mt. Hallasan National Park Visitor Center

The visitor center at Mt. Hallasan was established in April 2008 at the elevation of 980 m to provide natural and cultural values of the mountain. The center increases visitors' satisfaction by introducing Mt. Hallsan properly and aims to educate the scientific and cultural significance as a geoheritage.

The visitor center provides information desk and scientific displays including volcanic landforms and geology, legends and history related to the mountain, and ecological aspects though specimen, photographs and films. Also, multi-purposed media room and creative educational programs provide various educational opportunities. Mt. Hallasan meteorological station and art gallery are provided outside. Full time heritage interpreters (geoparkians) run educational programs or run guided tour programs (Figs. 8.2, 8.3, 8.4 and 8.5).

8.1.1.2 Folklore and Natural History Museum

Folklore and Natural History Museum was established in 1984 and is visited by more than 1 million people annually. The museum exhibits more than 4,000 specimens of geology, terrestrial and marine ecosystems, and Jeju culture. Especially, the museum runs socio-educational programs such as museum education for children, eco-education, science edu-

Fig. 8.5 Education program for elementary school students at the Eorimok visitor center

Fig. 8.7 Stone display at the Jeju Stone Culture Park

Fig. 8.6 Folklore and Natural History Museum

Fig. 8.8 Jeju Stone Culture Park surrounded by natural forests and cinder cones

cation, experience education for marine life, cultural events, traditional culture lectures. These programs are aimed for a variety of age levels from pre-school children to retired old people (Fig. 8.6).

8.1.1.3 Jeju Stone Culture Park

Jeju Stone Culture Park is all about stones in geography and geology, stones in history, stones in religion, stones in art, stones in sculpture, stones in houses and importance of stones in every sphere of life of the islanders of Jeju. The park (see the Box below) is a joint creation of the Jeju Island government and a local businessman. This truly remarkable creation is an integral part of the Jeju Island Geopark concept and again will be used to promote the Geopark. It will be a major education and information point for the Geopark.

The exhibition center in the park provides thorough displays dealing with the formation of Jeju Island related to volcanic activities and other geological aspects. Together with well prepared scientific exhibitions, stone galleries consisting of the wonderful display of erratic natural stones (volcanic rocks) deserve to be called as a 'artistic museum in perfect harmony with scientific information'. In the outside, evolutionary patterns about stone culture from prehistoric artifacts are well displayed surrounded by well preserved natural environments. Educational program for geology and stone culture are regularly carried out for elementary students and general public in the park (Figs. 8.7 and 8.8).

8.2 Educational Tourism

Fig. 8.9 World Heritage Center established in 2012

Fig. 8.11 Guided tourism in Hallasan

The center will provide all the necessary information on ecological values (Man and Biosphere) as well as world heritage values of the inscribed sites for educational purpose. The center will provide educational experience in some inscribed wild caves, education for the conservation of natural environments, scientific research, management and monitoring of the inscribed sites, and international corporation programs (Figs. 8.9 and 8.10).

8.2 Educational Tourism

Heritage interpreters and guides ('geoparkians') are a vital part of any geopark to assist visitors with information, to promote the geopark and to assist in the conservation of sites by monitoring visitor behavior and providing a regulatory presence if needed.

Since 2001 the Jeju Special Self-Governing Province has been training staff and volunteers to interpret Jeju's heritage for domestic and international visitors. Some 230 volunteer workers have been trained to give professional, informed explanations about geology, ecology, history and culture for visitors. In 2009, there are over 200 heritage interpreters and they are deployed to 26 sites throughout the Jeju Island Geopark. In addition, there are over 1,200 travel guides including general tour guides (ca. 350), cultural heritage interpreters (ca. 24), interpreters for international travelers (ca. 250), and Olle guides. These guides are regularly provided for educational programs by museums, local universities and colleges, and research institutes.

In addition, the organization was made by heritage interpreters since 2005 and more than 170 members are actively involved. The organization hosts workshops and carry out field trips for better understanding of volcanic landforms. In addition, it invites various experts in folklore, ecology and history to organize special lectures and has published six reports on Jeju cultural heritages until 2011 (Figs. 8.11, 8.12 and 8.13).

Fig. 8.10 Experience corner for a lava tube cave at the Promotion Hall in the World Heritage Center

---

Jeju Stone Culture Park
- Work with the environment as our priority
- Work with Jeju's identity, local character, and artistry in our mind.
- Work with the past, present and future of stone culture of Jeju in our mind.

8.1.1.4 World Heritage Center

The Jeju Island World Heritage property, which was inscribed by UNESCO in 2007, is constructing a major visitor center at Geomunoreum (the source of the lava flow that created the Manjang Cave Geosite). The provincial government of the Jeju Special Self-Governing Province will seek to ensure that the role and features of the Jeju Island Geopark are adequately presented in this centre.

Guided Tour Program Proposed for the Sanbangsan and Yongmeori Geosites
Title
- From the Sky to the Sea, a Vertical Complex Geosite
Main Themes
- Geological features from Sanbangsan Lava Dome to the Yongmeori Tuff Ring
- Cultural and historical features of the site

Methods and Target Audiences
- Explanation by guides
- Use of multi-media devices (PDA, PMP etc)
- Target audiences: Children, students, adults, retirees and international tourists

Visiting Route
- Sanbanggulsa Temple–Sanbangsan–Hamel Museum

Contents
- Sanbangsan (30 min travel time from the visitor center): (1) Formation and geological features of the mountain, (2) Legends of the mountain, and (3) Mountain ecosystem
- Sanbanggulsa Temple (20 min): Buddhism on Jeju Island and legends of the temple
- Hamel Museum (30 min): (1) Hendrick Hamel and VOC (Dutch East India Company), and foreign visitors to Jeju, and (2) Relationships between Korea and the other countries
- Yongmeori Coast (20 min): (1) Basic geological knowledge and formation.

Fig. 8.12 Field survey with history expert. (Photo by Jeju Culture and Tourism Interpreters' Organization)

Fig. 8.13 Field survey to Mt. Baegdu in China

There are many opportunities for enhanced geotourism on Jeju Island to support the geopark concept, to develop local business prospects and to contribute the overall sustainable development of the Island. These include the development of guided tours focused on the geosites and Jeju's geoheritage and cultural sites on foot, by bicycle or motor vehicle.

KIGAM's (Korea Institute of Geoscience and Mineral Resources) recent "Guidebook for a Geological Tour of Jeju Island" (Park 2006) is a very valuable addition to the Island's large collection of tourist information books, maps and brochures. Although written mainly in Korean it has sufficient English comments to make it valuable for those visitors with a basic command of English. Consideration should be given to translating the book, which was published with the assistance of the Jeju Development Center, into Japanese, Chinese and other appropriate languages. The book will become the basis for successful Jeju Island Geopark activities.

One initiative taken by the local community is the development of the Olle walking/cycling track that runs around the southern coastline from the Seongsan Ilchulbong to the Suweolbong Geosites (Figure 9.6). Although this initiative was not undertaken with the geopark concept in mind it forms an excellent pattern for future activities of its kind. It is made up of 15 separate, but joined, sectors with an additional one under development on Udo (U Island). There are excellent guides in Korean and English to the Olle track system. However, the geoheritage aspects of the Olle system need to be augmented with better training of guides and establishing linkages between the trail system, the geosites and other natural heritage features along the routes. The Jeju government has supported this community initiative with grants totaling more than US$ 650,000. The number of users from 2007 until October 2009 is estimated to be more than 200,000 and income generated, based on surveys by government and the Olle organizers, is said to be US$ 11.4 million. Twenty-five Olle courses were developed up to 2012, and improvements to rest areas, accommodation facilities, signboards and restoration of roads have been made (see the next chapter for more detail).

A further boost to geotourism on Jeju Island will be a Korean Geoparks Network with additional Korean geoparks based on the Jeju Island Geopark model. Such a network will

8.2 Educational Tourism

Fig. 8.14 "Ami's Dream", a book published for children for understanding natural heritages on Jeju Island

enhance Korean geotourism by cross-linking and promoting geoparks across the country. It will also enhance the understanding of the Korean people in geological matters generally as these are not comprehensively taught in the Korean school system (Figs. 8.14, 8.15 and 8.16).

The Olle trail system (see Chap. 9.3.3.1) stretches for more than 200 km along the eastern, southern and western coasts of Jeju and links many of the geosites. Much of this system can be covered by bike and the main island road has cycle-lanes. There is an extensive system of walking trails linking the sites in the Hallasan Geosite Cluster as well as developed walkways at Seongsan Ilchulbong. The significant Gotjawal forests growing on aa (clinker) lava flows are serviced by an 8 km walking trail at Geomunoreum (the site of the World Heritage Center). These major trails are supplemented by many other walking and cycling routes.

The Jeju Olle has attracted local residents by establishing partnerships between local village and production companies and also by providing an internet website for selling lo-

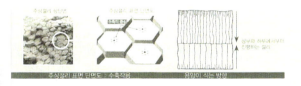

Fig. 8.15 A book published for the geosite of the Jungmun Daepo Columnar-jointed Lava

Fig. 8.16 A fieldtrip to the geosite (the Seoguipo Formation) by geosciences high school teachers

Fig. 8.17 A friendship route was opened in April 2011 between Giant Causeway (UK) and Jeju Olle. (Photo provided by Jeju Olle)

cal agricultural products. It developed educational program (Olle Academy) to educate Olle guides about local history, culture and ecology. About 400 Olle guides (Ollegil-dongmu) have been educated and over 50 guides are actively working (Fig. 8.17).

Economy and Development of Sustainable Tourism

9

The Jeju Island Geopark has a fundamental objective to play an active role in the economic development of the island through enhancement of an identity and mission to link the geological and other heritage with geotourism and public awareness and education. This will require the development with strong and effective links between: (1) the Jeju Island Department of Geoparks, (2) the Jeju Volcanic Island and Lava Tubes World Heritage management office, (3) the Jeju Development Center, (4) other branches of the government of the Jeju Special Self-Governing Province, (5) the nine geosites, (6) other natural and cultural heritage sites and their management, (7) all elements of the tourism industry including transport operators, (8) non-government organizations, (9) schools and academic institutions, and (10) interested individuals.

A Jeju Island Geopark 'pass' will be developed so that a single, easily obtainable, ticket will provide access to all of the current and future geosites (all the geosites, except the Seogwipo fossils and the Suweolbong Tuff Ring, currently charge at least a parking fee).

9.1 Economy

Jeju Island is a province of the Republic of Korea. The Self-Governing Province status allows the Island to be a "free international city" with no visas required for direct entry and no taxes are levied. It has a dynamic and growing economy largely dependent on the Tertiary Industry Sector which includes tourism, insurance, government, banking, retail and education activities. Since 1950, Jeju government has promoted tourism industry as a major tool for economic development. As a result, various cultural and touristic infrastructures dealing with history, culture, and nature are present on Jeju Island.

The tourism industry is extremely active and is served by up to 23 airlines daily with about 240 and 38 domestic and international flights, respectively each day. Annual aircraft movements in 2011 amounted to about 117,000 flights carrying 17 million passengers. Jeju International Airport is the second largest in Korea (after Incheon International Airport) and ranked above the Gimpo Domestic Airport in Seoul since 2009. Jeju is also served by daily ferry services from eight ports on the Korean Peninsula carrying nearly 3 million passengers a year.

At present, the tourist industry makes up 70 % of Jeju economy and has been growing continuously. For example, the number of tourists has increased by 4.5 % annually since 2003. Especially international tourists have increased since the World Heritage inscription in 2007 and the Global Geoparks Network endorsement in 2010. In 2009, the number of tourists who visited Jeju Island was 6,523,938 and international visitors were 632,354 making up 9.3 % of the whole tourists. International visitors have increased from about 220,000 to 630,000 probably possibly due to several reasons: (1) expansion of the entry without visa, (2) designation of International Convention City, (3) inscription of World Natural Heritage, (4) Endorsement of Safe City by WHO, (4) development of the Olle trails, (5) establishment of Jeju Tourism Corporation, (6) revitalization of direct international flight routes, and (7) active marketing by Jeju government. A half of tourists are visitors for travel and sightseeing, and recently visitors for business and conference have increased. Recent statistics revealed that there are 149 hotels and lodging places including 12,092 rooms. There are 61 international and domestic travel agencies, 91 international travel agencies and 496 domestic travel agencies. There are about 50 museums, art galleries and exhibition centers, and public and private tourist attraction sites are 50 and 79, respectively.

The Island is served by an efficient road system, and buses and taxis are ubiquitous, frequent and cheap by international standards. The nine geosites within the Jeju Island Geopark are described above in Sect. 2.2. They are among the premier tourist sites on this tourist island and already receive considerable numbers of visitors after the World Heritage inscription by UNESCO. All but one of the nine geosites can readily be reached by bus. The same can be said for most of the geosites identified for future addition to the Jeju Island Geopark network (Fig. 9.1).

Fig. 9.1 Many tourists at the Seongsan Ilchulbong Tuff Cone geosite

Fig. 9.2 Jeju Airport in 1960s. Japanese tourists got off Japanese in the rain. Photo from "History of Jeju Learning by its Photos" (Jeju Special Self-go)

9.2 Sustainable Development

9.2.1 History of Tourism on Jeju Island

In 1950s, Jeju government began to be interested in tourism industry. In 1956, Jeju government made a catchphrase "The island of *Samda*, *Sammu*, and *Sambo*" meaning "The island of *Three plentys* and *Three absences*", and established the first tourist information center on Jeju Island (see Chap. 4). In 1960s, basic infrastructure for tour activities has been provided and Jeju tourism has started. Especially, Jeju Tourism Association was established in 1962, and official tour guide system was adopted in 1966, which was the first system in Korea.

In 1970, Jeju tourism has been developed by central government, and tourists to Jeju Island exceeded 500,000 people in 1977. In 1980s, Jeju Island became one of the most popular sites in Korea as a honeymoon place. It is also the tourist attraction place recommended by children to their parents because it is common for children to support the expense for the travel of their parents when they are older than 60. Since 1990s, comprehensive development plan was initiated and 16 tour sites were designated and developed. The first inscription of World Natural Heritage by UNESCO in 2007 has increased international recognition (Figs. 9.2 and 9.3).

9.2.2 Development of Sustainable Tourism

The Jeju Special Self-Governing Provincial government through agencies such as the Jeju Development Centre and with the assistance of business has plans to dramatically increase the facilities and functions of tourism on the Island. Limits on water supply, power and so on means that development has to be sustainable for the Island to have a viable future. Plans include: (1) expansion of Jeju International

Fig. 9.3 A ceremony held in Jeju International Airport for the opening of flight route (Jeju–Busan–Osaka) by Korean Airline in October 1969. Since then, Jeju–Tokyo route was developed in October 1985, and Jeju–Nagoya route in March 1988. Photo from "History of Jeju Learning by its Photos" (Jeju Special Self-go)

Airport capacity from 11 to 23 million passengers annually, (2) conducting a feasibility study for a second airport, (3) expansion of berth facilities in Jeju Port from 2.55 to > 3.36 km to cater for ships of up to 80,000 tons, (4) road improvement and expansion and establishment, (5) augmentation of the power supply from 145,000 to 245,000 kW, in part from renewable sources, (6) augmentation of the sewage disposal network from 67 to > 90 % of the Island, (7) establishment of the World Heritage Center and the Geopark facilities, and (8) establishment of many new large-scale tourism ventures.

The tourism investment is expected to create 4,000 new jobs at a time when public institutions are scaling back new hiring or cutting their workforces. It is estimated that, when all tourism projects are completed, that some 18,000 new jobs for the island's residents would be created.

A budget for the future operation and development of the Jeju Island Geopark has been approved by the Jeju Special Self-Governing Province. It includes funds for management of the geosites, scientific research, education and promotion including input into community groups and the training of heritage interpreters as well as capital works.

9.3 Socioeconomic Development

9.3.1 Jeju Tourism Organization and Jeju Special Self-Governing Provincial Tourism Association

Jeju Tourism Organization, which was established in 2008 for local economic development, the promotion of tourism industry, and welfare promotion, developed the partnership with Jeju Special Self-Governing Provincial Tourism Association established in 1962. This partnership promotes efficient geopark activities and economic development. Also, this promotes networking between local tourism industries. Jeju Special Self-Governing Provincial Tourism Association has 45 different types of business and 658 enrolled members.

9.3.2 Geomunoreum Trail Run by Local Residents

In July 2007, an international trekking competition was held for 2 months in the Seonheul village near the Geomunoreum cinder cone from which the Geomunoreum Lava Tube System was formed. The competition was for the first anniversary of the inscription of "the Jeju Volcanic Landforms and Lava Tubes" as a World Natural Heritage by UNESCO. This competition was a temporary event to promote the World Heritage by Jeju Special Self-Governing Provincial Government, local residents, local press and travel agencies. During the competition, as the trekking course at the Geomunoreum became well known and more visitors and local residents visited the course, this trekking course became one of the most favorite trekking sites on Jeju Island. As a result, this trekking course was opened to the public as a commercial one after September 2008.

Along the trekking course, some infrastructure was built at a few sites for the conservation of natural environment and the safety of visitors. Because local residents was involved as natural heritage interpreters since the first International Trekking Competition in 2008, it did not take long for this site to be run as commercial permanent trekking place managed by the Geomunoreum Trekking Information Center.

Fig. 9.4 Guided tourism in the future geosite, Geomunoreum. The guided tour system is run by local residents

Since 2009, local residents near the Geomunoreum have managed the site entrusted by the Department of World Heritage Management in Jeju government.

The trekking course is maintained with great care by limiting the maximum number of visitors up to 100 people per day during weekdays and up to 300 people during weekends. As the number of visitors increased, the maximum capacity of visitors was determined to be 300 people per day. Visitors can come only by reservation. Visitors are guided by voluntary heritage interpreters every 30 min. The heritage interpreters as well as safety guides and heritage managers are educated by the programs provided by the Department of World Heritage Management in Jeju government.

Souvenir shop is run by local villagers and various items such as trekking equipments, local agricultural and industrial products are being sold. In addition, various camping programs and farming experience programs planned and will be provided near the site in the near future for better socioeconomic and sustainable development (Fig. 9.4).

9.3.3 Partnership

Tertiary industries comprise about 78 % of Jeju economy, and more than 70 % of the population on Jeju Island is involved. In the Jeju Island Geopark, there are 129 museums and tourist attraction sites, and 50 of them are run by the Jeju Special Self-Governing Province. To connect tour infrastructures organically and to utilize them, the Jeju Island Geopark has developed partnerships with a private organization and tourism and science research institutes. The partners are supposed to keep the code of practice (see the box below). A few partners are introduced below.

9.3.3.1 Olle Walking Track System

Olle (Ole) is the Jeju word for a narrow pathway that is connected from the street to the front gate of a house. Hence, "Olle" is a path that comes out from a secret room to an open space and a gateway to the world. If the road is connected, it

Fig. 9.5 Education of the Olle Academy at the resort place along the Olle trail. (Photo from http://www.jejuolle.org)

is linked to the whole island and the rest of the world as well. Also, "Olle" has the same sound as "Would you come?" in Korean, so Jeju's "Olle" sounds the same as 'Would you come to Jeju?'

The first trail route was opened to the public in September, 2007. Since then, the Jeju Olle exploration team has created a combined total of 367 km of walking trails on Jeju Island. In May 2011, 23 trail routes have been opened to walkers and the trail exploration team is working on new routes. Jeju Olle hopes all walkers who explore the Jeju Olle trail routes gain "peace, happiness and healing" on the road. We've traveled at high speed so far, so it is time to give your soul a moment's rest. The information can be found from http://www.jejuolle.org/eng/.

Jeju Olle opened information desk at the Jeju International Airport and signed the partnership with local hotels.

Development of the Olle souvenirs and Olle passport system have been favored by Korean and international visitors. The Jeju Olle organized the World Trail Conference in October 2010 and has developed 'friendship trails' with Switzerland and United Kingdom (Giant Causeway) (Figs. 9.5 and 9.6).

> **Jeju Island Geopark Cooperative Partner CODE OF PRACTICE**
>
> The Jeju Island Geopark Committee and Jejuolle are committed to observe and follow the code below for the promotion of the Jeju Island Geopark.
>
> 1. Jejuolle (as a member of the Jeju Island Geopark Partnership Association) will cooperate with the Jeju Island Geopark Committee and the committee will support the work of the members.
> 2. Jejuolle will promote the Jeju Island Geopark through geological programs by using the official logo of the Jeju Island Geopark and will not provide it to any third party based its own discretion or without permission.
> 3. Jejuolle will not develop or sell products made with local geological materials, in order to preserve Jeju's clean environmental resources.
> 4. Jejuolle will maintain the credibility of the Jeju Island Geopark by avoiding poor business practices including overcharges, poor customer service, and anything else considered inappropriate.
>
> June (24th), 2010
> **Suh Myung-Sook CEO, Jejuolle**

Fig. 9.6 Olle trails on Jeju Island

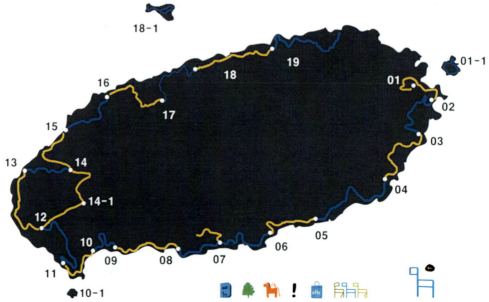

9.3 Socioeconomic Development

Fig. 9.7 Songaksan tuff cone and cinder cone viewed from the sea

Fig. 9.8 A Green Cruise. Sambangsan and Yongmeori geosites are seen behind

9.3.3.2 'Gal-jeung-I': Local Handicraft Company

By using the traditional natural dying technique of Jeju Island, the company developed various types of clothes and accessories. It also runs natural dye experience program in harmony with the concept of natural conservation. Due to the partnership with the Jeju Island Geopark, it tries to promote the geopark outside Jeju Island.

9.3.3.3 Green Cruise

The Green Cruise is a private company which developed sightseeing courses on a tour boat. The main tour course includes Mt. Sanbang and Yongmeori geosites as well as the fascinating scenery of Mt. Songak. Through the partnership with the Jeju Island Geopark, the tour will include geological interpretation by developing various contents (Figs. 9.7 and 9.8).

Management Plan

10

10.1 Purpose of the Management Plan

The purpose of the Management Plan is to provide a framework for the care, control and management of nine geosites as the focal points of the Jeju Island Geopark. The plan supports protection of the geosites, development of geotourism and the ongoing economic development of the Jeju Island Geopark in a sustainable manner. The Plan has three main elements as follows: (1) A management structure to ensure that the geosites are managed sustainably and in a coordinated manner throughout the Jeju Island Geopark in association with other natural, cultural, social and economic sites and values; (2) Site management and monitoring concepts to ensure that the natural and cultural values are protected and enhanced for this and future generations; and (3) The establishment of linkages and systems to promote and develop education, tourism and research can work together to promote and enhance an understanding of geological heritage whilst contributing to Jeju's economy and social well-being.

This is an interim Management Plan for the Jeju Island National Geopark. It will generally follow the prescriptions of the Jeju World Natural Heritage Revised Management Plan (2006) which is the formal Plan for the three World Heritage listed geosites and Jeju Geopark Management Plan (2009, 2011).

The Plan will need to address the issues of establishing a Jeju Island Geopark Regional identity; increasing effective information transfer in the product region, facilitating effective community involvement in tourism, increasing awareness of the Jeju Island Geopark in source markets; identifying what is needed to enhance Jeju Island Geopark attractions; identifying what is needed to upgrade access to the Geopark attractions; and identifying what is needed to enhance visitor accommodation and tourism support infrastructure. The plan will also need to identify the location and function of interpretive centers, locations for onsite interpretation, the designation of touring routes and locations requiring significant signage.

10.2 Legal Basis for Protection and Management

10.2.1 Legal Basis for Protection

The protected areas are protected and managed under the national laws of the Republic of Korea and by the local governments under many specific laws. First of all the Constitution of the Republic of Korea defines the protection and transmission of traditional and national culture as the responsibilities of the country (The Constitution of the Republic of Korea, Article 9). Therefore the protection of cultural heritage is the fundamental responsibility of the nation and the law that specifies this is the Cultural Heritage Protection Act 2007.

In addition to the Cultural Heritage Protection Act and the Jeju World Heritage Revised Management Plan the followings have relevance to the management of natural heritage on Jeju; Natural Parks Act (1980), Special Act on Jeju Free International City (2002), Highlands Management Act (2005), National Land Planning and Utilization Act (2002), Coastal Zone Management Act (1999), Environment Conservation Management Plan for Jeju Province, Jeju Island Biosphere Reserve Management Plan, and the Hallasan National Park Management Plan.

As all the nine geosites are Korean Natural Monuments, the legal basis for protection of geosites ultimately lies with the national Cultural Heritage Protection Act 2007. These Acts are supplemented by regulations, by provincial ordinances and regulations, and by Administrative Directives from the Cultural Heritage Administration. The legal status of each of the nine geosites is shown in the following table (Table 10.1).

Although the nominated properties are being conserved and managed as natural monuments pursuant to the Cultural Properties Protection Act and as national or county parks pursuant to the Natural Parks Act, the properties, together with Hallasan Natural Reserve require special care and attention as they contain a high concentration of valuable plant, animal, mineral resources and landforms. The general provisions of the statutes are as follows:

K. S. Woo et al., *Jeju Island Geopark—A Volcanic Wonder of Korea*, Geoparks of the World,
DOI 10.1007/978-3-642-20564-4_10, © Springer Verlag Berlin Heidelberg 2013

Table 10.1 Tenure and status of the nine geosites within the Jeju Island Geopark

Geosite	World Heritage	Man and the Biosphere	National monument number	Other protected area status	Buffer zone present
Hallasan Geosite Cluster	Yes	Yes	182	National park	Yes
Manjang Cave	Yes	No	98		Yes
Cheonjiyeon Waterfall	No	Yes	378 & 379		Yes
Daepodong Columnar Joints	No	No	443	Protected coastal zone	Yes
Seogwipo Formation	No	No	195	Protected coastal zone	Yes
Seongsan Ilchulbong Tuff Cone	Yes	No	420	Nature reserve	Yes
Sanbangsan Lava Dome	No	No	376		Yes
Yongmeori Tuff Ring	No	No	526	Protected coastal zone	No
Suweolbong Tuff Ring	No	No	513	Protected coastal zone	Yes

10.2.1.1 Natural Monument

Resources designated by the Cultural Properties Protection Act, based on deliberations by the Cultural Properties Committee, to conserve natural heritage with scientific values that are rare by domestic and international standards. The Cultural Properties Protection Act defines them as animals (including their native habitats, breeding ground and refuges), plants (including their native habitats), minerals, lava tubes, geological features, biological formations and special natural phenomena containing rich historic, scenic or scientific values.

Hallasan Natural Reserve, Geomunoreum Lava Tube System and Seongsan Ilchulbong Tuff Cone are all designated as Natural Monuments, based on the Cultural Properties Protection Act. According to this act, any conservation, management or utilization of cultural properties are performed on the basic principle of maintaining their original forms. Based on the above act, permission from the Administrator of the Cultural Heritage Administration is required for the followings (Article 20): (1) capturing or collecting animals, plants or minerals within an area designated or provisionally designated as a scenic area or a natural monument, or within its protected zone, or carrying them out of such an area or zone; (2) taking any rubbing, or photoprinting of State-designated cultural properties, or making a film of them in such a manner that may affect their preservation; and (3) any acts as prescribed by the Ordinance of the Ministry of Culture and Tourism, which are such acts as altering the current status (including the act of sampling or stuffing the natural monuments) of the State-designated cultural properties (including the protected objects and protected zones, or as affecting their preservation.

For opening to the public of State-designated Cultural Properties (Article 33), when deemed necessary to preserve the State-designated cultural properties and to prevent them from being damaged, the Administrator of the Cultural Heritage Administration may set limits to any opening to the public of the whole or part of relevant cultural properties.

10.2.1.2 National and County Parks

A national park is an area that is representative of the nature, ecosystems and culture of Korea and designated and managed by the Minister of Environment pursuant to the Natural Parks Act. A country park is also an area that is representative of the natural ecosystems or scenery of a city or county, and designated and managed by the mayor of the relevant city or county pursuant to the Natural Parks Act.

Hallasan Natural Reserve and Seonsan Ilchunbong Tuff Cone are designated as National Park and County Park respectively under the Natural Parks Act. According to this Act, the State, local governments, persons who undertake park projects or manage park facilities, persons who occupy or use the natural parks, persons who enter the natural parks and persons who reside in the natural parks shall do everything they can in order to protect the natural parks, and maintain and restore order therein. The State and local governments shall designate areas characterized by picturesque scenery and excellent natural ecosystems as natural parks, and conserve and manage such designated natural parks sustainably.

Based on the above act, permission from the Minister of Environment is required for the followings (Article 23). Any person, who intends to perform an act in the park, falling under each of the following subparagraphs will obtain permission from a park management authority under the conditions as prescribed by the Presidential Decree: (1) newly building, extending, remodeling, reconstructing or relocating structures as well as any other installations; (2) mining minerals and collecting earth, stones and aggregates; (3) clearing land as well as any other act of altering the form and quality of land; (4) cutting timbers or collecting wild plants; (5) putting cattle out on pasture; and (6) damaging the scenery and changing the purpose of use of structures which is feared to impede the conservation and management of a natural park.

Under the above act, following undertakings are strictly prohibited (Article 27): (1) disrupting the present state of any natural park or damaging park facilities; (2) damaging trees and catching wild animals; (3) performing a commercial transaction outside a designated place; (4) camping outside a designated place; (5) parking outside a designated place; (6) cooking outside a designated place; and (7) littering.

The park management authority may, if it is deemed necessary to protect a natural park, restore damaged nature, ensure the safety of persons entering such natural park and enhance the public interest, limit or prohibit access to a certain area of such natural park for persons and vehicles for a fixed period.

10.2.1.3 Natural Reserve

A nature reserve is a type of natural monument with rich resources that require protection. More specifically, it is an area that represents the interaction between man and nature with a repository of diverse cultural, historic, scenic, geological and biological evolution processes.

The local governments of Jeju-do (Jeju Province) shall endeavor to enact their basic environment ordinances and to formulate and implement their basic environment preservation plans including matters set forth in the following subparagraphs in order to systematically preserve and manage the natural environment and to ensure that their residents may live healthy and comfortable lives in a pleasant and agreeable natural environment: (1) presentation of the objectives and direction of environmental conservation; (2) analysis of the features of the regional environment and future prospects therefore; (3) plan for the preservation and restoration of the natural environment and ecosystem, etc.; (4) matters concerning the preservation and management of urban and natural sceneries; and (5) matters concerning the management of the Jeju Island Biosphere Reserve designated by UNESCO.

There shall not be permitted within the core zone such acts as the construction of a building, setting up of a structure and other facilities, alteration to land form and nature, partition of land, public waters reclamation, logging, exploitation of soil and stones, construction of new roads, or any other acts similar thereto, which may be contrary to the purposes of the designation of such an area: Provided, that this shall not apply where acts which fall under any of the following subparagraphs have been permitted by the Governor (Artic-

le 27): (1) construction works carried out by the State or local governments, such as works for paths up to mountains, promenades, forest paths, roads, public lavatories, pavilions, meteorological observation facilities and park facilities under the Natural Parks Act; (2) afforestation projects carried out as a forest management plan under the Forestry Act, which is not accompanied by logging or alteration to land form and nature; (3) activities performed for the purposes of academic research and study; (4) extension or alteration to the existing buildings in the precincts of a religious establishment which was constructed before the designation of the Absolute Preserved Area; and (5) such other activities as determined by the Jeju Provincial Ordinance insofar as they cause neither damages nor alteration to natural resources.

10.2.2 Legal Basis for Management

Since the Jeju Island Geopark was endorsed in 2010, local governments have shown interests to become a member of the Global Network of National Geoparks. Because aspiring geoparks needs financial support, the Ministry of Environment established the a law on geoparks. Even though it is the revised pre-existing legislature for national parks, it can become the basis to support aspiring geoparks financially from the Korea Government. It is the revised one of the National Parks Act (1980) dealing from protection and management of geosites, financial support for national geoparks, endorsement standard of national geoparks to the training and role of geoheritage interpreters (geoparkians). The national geoparks in Korea will be endorsed by the Minister at the Ministry of Environment after submission of proposals by local governments. Also, as in the case of national geoparks of the GGN member, endorsed national geoparks should be revalidated every 4 years. This system will promote active development of national geoparks in the near future.

10.3 Management Structure

The Jeju Island Geopark and its constituent geosites must operate within the existing framework of the government of Jeju Special Self-Governing Province, applicable Korean laws and regulations as well as the requirements for the management of the Jeju Volcanic Island and Lava Tubes World Heritage property as set out in the UNESCO World Heritage Convention and in the Revised Management Plan. Should the Jeju Island Geopark be accepted into the Global Geoparks Network the principles and practices of the Network would be applied by the Jeju Special Self-Governing Province.

The Plan sets out the functions of a management structure for the Jeju Island Geopark in association with the management of the World Heritage property and with the existing Jeju Island Department of Monuments which will continue to have responsibility for natural and cultural monuments outside the World Heritage and the Jeju Island Geosites. Clearly close cooperation and liaison will be needed between the three departments as there is, inevitably, overlap between their roles and responsibilities. The complete structure will evolve and expand through time as the Geopark evolves and is expanded through the additional geosites to be reviewed and added as outlined in Sect. 2.3.

The three main branches of the Department of Geoparks and their responsibilities are as follows. A promotion and tourism branch which will: (1) develop marketing and promotional events in cooperation with the community and tourism-oriented businesses; (2) cross-promote, where appropriate, related tourism activities; (3) ensure that the activities of the Jeju Island Geopark are in accord with the principles and practices of the Global Geoparks Network; and (4) promote the development of a Korean Geoparks Network and establish links with the Asian Pacific Geoheritage and Geoparks Network. A research, conservation and education branch which will: (1) assess the sites foreshadowed as additional geosites for inclusion in the Jeju Island Geopark; (2) monitor the condition of the geosites and develop maintenance and rehabilitation programs; (3) foster appropriate research on the geosites; (4) review research and other information and convey this to operational staff and volunteer 'georangers' to ensure that the Geopark 'message' is conveyed effectively to visitors; (5) develop and oversee education programs and materials; and (6) conduct training activities for staff, volunteers and local communities. A management branch which will: (1) oversee the activities of the four geosite visitor centers and the five visitor points; (2) deliver educational activities; (3) conduct special events and promotions; (4) carry out maintenance as appropriate; (5) execute the departmental administrative functions.

The Jeju Island Geopark Council (JIGC) was established, and the council has two branches (Research and planning branch and Tourism and promotion branch). Korea central government, Jeju government, research institute, non-governmental organization, local community council, heritage interpreters' association and schools are involved in the JIGC. It is responsible for the management of the Jeju Island Geopark together the Department of World Natural Heritage Management. The department is also responsible for the management of the "Man and Biosphere' on Jeju Island. Mt. Halla Research Institute which belongs to the Jeju Special Self-Governing Province has been involved with biological research and monitoring of Mt. Halla. It is hoped that the institute will broaden its research scope more toward geological aspects. Since Mt. Halla is also a national park. Park office of the Mt. Halla National Parks manages tracking courses and visitors.

10.4 Potential Pressures on the Geosites

The Jeju Island Geopark is subject to a wide range of pressures being on a relatively small and heavily populated island with a vibrant and developing economy. The legal protection given most of the geosites largely isolates them from direct physical impacts, but off-site disturbances may dramatically impinge upon the aesthetics of the geosites. This is especially the case for the coastal sites. The natural monuments are Designated Tourist Zones and thus outside pressures are subject to strict planning controls.

The Department of Geoparks will liaise closely with the Jeju Government planning authorities and with developers to ensure that impacts on the geosites are minimized consistent with the desire to support sustainable economic development.

10.4.1 Natural Processes

Natural environmental processes, including those processes operating on geological timescales, produced the landscapes of Jeju Island and the nine geosites. The most significant of these processes, other than the original volcanic activity, is coastal erosion which directly affects most of the nine geosites (the exceptions being Hallasan, Manjang Cave and the Cheonjiyeon Waterfall). The rates of coastal erosion are such that the six affected geosites will be present for millennia with erosion continuing to expose their salient features.

Manjang Cave has the ever-present possibility (which cannot be predicted) of roof collapse to produce another entrance in addition to the three entrances produced by natural collapse. The Cheonjiyeon Waterfall will continue to retreat at geologically appropriate rates unless human disturbance upstream alters the flow regime. This appears unlikely.

10.4.1.1 Climate Change
Climate change scenarios for Korea, building on observed increases in temperature, sea level rise and increased storminess as well as climatic modeling, suggest that sea level rise and increased frequency and intensity of typhoons and other extreme rainfall events can be confidently expected. These changes will inevitably accelerate the impacts of the natural environmental processes discussed above. Sea level rises will accelerate the rates of coastal erosion and cannot feasibly be ameliorated. Reduced snowpack may reduce the rates of groundwater recharge.

10.4.1.2 Urbanization and Infrastructure Development
Given the development pressures consequent on a growing and dynamic economy the potential impacts on the geosites are large but, because of the legal protection afforded to most

10.6 Future Action Plans

of the geosites, the impacts will largely be confined to offsite visual impacts. The currently unprotected Yongmeori and Suweolbong Tuff Ring Geosites are relatively remote from urban pressures. The Department of Geoparks will liaise closely with the Jeju Government planning authorities and with developers to ensure that impacts on the geosites are minimized. The Designated Tourist Zones provide protection from outside pressures and are subject to strict planning controls.

10.4.1.3 Quarrying

As quarrying is strictly regulated under the various statutes and plans pertaining to the protected areas this will not be an issue except potentially (in the short term) for the Suweolbong and Yongmeori Geosites. A greater danger lies with extractive industries altering the viewscapes from a geosite in an inappropriate way. The Department of Geoparks will liaise closely with the Jeju Government planning authorities and with developers to ensure that impacts on the geosites are minimized.

The culturally significant carved stone figures known as Dolhareubang (Jeju Province Folklore Material Number 2) are traditionally life-sized or larger figures, but over recent decades small Dolhareubang 15–20 cm in size (and larger) have been produced as popular souvenirs for the tourist trade. Many other types of stone carvings are produced for the domestic and tourism trade. Stone to supply this industry can be sourced from sites well removed from the current and future geosites. The production and sale of stone carvings in the Jeju Island Geopark is an important traditional industry and will continue within Geopark but outside the current and future geosites.

10.4.1.4 Groundwater Extraction

As none of the geosites rely on hydrostatic support groundwater extraction should not be a threatening process except perhaps for the Cheonjiyeon Waterfall Geosite where there might be a potential to alter flow regimes. However, all of their catchment, surface and underground, is within protected areas.

10.4.1.5 Agriculture and Forestry

Agriculture and forestry are very long-term traditional land uses on Jeju and these activities form the backdrop to many of the geosites. Impacts from these activities are chiefly visual although in the case of the Yongmeori and Suweolbong Geosites there may be minor interactions between visitors and rural activities. Additional protection will be provided by the Department of Geoparks.

10.4.1.6 Pollution

Gross air or water pollution is unlikely to affect the geosites although a marine oil spill might impact negatively on the coastal geosites in the short term. Of far more importance is littering especially where the geosites are not provided with visitor facilities or are regularly patrolled as in the case in rural areas and on the margins of some geosites. Special attention will be paid to remove trash with in geosites. Seaborne litter is a major issue on some coastal sites but this is beyond remediation by Jeju authorities.

10.4.1.7 Visitor Use

As discussed below (Sect. 10.5, Current management practices and facilities) most of the geosites have at least walking trails and viewing platforms to guide visitors and to protect the sites. In some cases, as at Daepodong, Cheonjiyeon and in Manjang Cave these are very sophisticated and well-maintained structures. In any case, the actual geological features exposed and displayed in the geosites are robust or inaccessible. Collecting of geologic materials is strictly forbidden and the regulations enforced. The fossils at Seogwipo will be protected by patrolling and by enlisting the local community in enforcing the prohibitions on collecting. Graffiti does to appear not be an issue. As mentioned above, littering can be a problem. Recycling opportunities are widespread on the Island. The Geopark will encourage recycling and promote anti-littering campaigns.

The Department of Geoparks will be developing monitoring and maintenance programs and codes of behavior to minimize visitor impacts (see Sect. 10.7, Future Action Plans, below).

10.5 Current Management Practices and Facilities

The array of current management practices and facilities in the nine geosites ranges from outstanding (in the World Heritage sites) to minimal (at Yongmeori and Suweolbong). A summary of these is provided in Table 10.2 above. The more heavily utilized areas are regularly patrolled and maintained, visitor activities and pressures monitored and volunteer georangers will provide some measure of interpretation and protection over and above that provided by interpretation panels and paid staff. The Department of Geoparks will continue to develop and maintain visitor facilities as discussed in Chap. 9 and Sect. 10.6, Future Action Plans, below.

10.6 Future Action Plans

A series of Action Plans, including facility development, will be developed by the Department of Geoparks and relevant specialists to formalize and develop the Jeju Island Geopark. The Action Plans will have a life of five years from date of their adoption. They will form part of the overall Manage-

Table 10.2 Visitor facilities and practices in the nine geosites

Geosite	Walking trails*+/viewing platforms	Interpretation panels present	Guided tours available	Educational groups specifically catered for	Toilets/food outlets
Hallasan Geosite Cluster	Yes/yes*	Yes	No	Yes	Yes/yes
Manjang Cave	Yes/yes	Yes	Yes	Yes	Yes/yes
Cheonjiyeon Waterfall	Yes*+/yes	Yes	Yes	Yes	Yes/yes
Daepodong Columnar Joints	Yes*+/yes	Yes	Yes	Yes	Yes/yes
Seogwipo Formation	No	Yes	By arrangement	By arrangement	Yes
Seongsan Ilchulbong Tuff Cone	Yes*+/yes	Yes	Yes	Yes	Yes/yes
Sanbangsan Lava Dome	Yes+/yes	Yes	No	Yes	Yes/yes
Yongmeori Tuff Ring	Yes+/no	No	No	No	No/no
Suweolbong Tuff Ring	Yes+/no	No	Yes	Yes	No/no

*Wheelchair accessible in whole or part
+Part of the Olle trail system (see Sect. 5.2 above)

ment Plan for the Jeju Island Geopark. The Action Plans will include:

10.6.1 Developing Visitor Centers

A series of four visitor centers and five visitor points are proposed at the geosites as well as two centralized information centers in Jeju and Seogwipo cities as discussed in Sect. 8.1, Education and Promotion, above. This Action Plan will provide the planning and scheduling for the construction and operation of these sites.

10.6.2 Promotion of Geopark and Geosites

This Action Plan will develop a system of promoting the tourism potential of the Jeju Island Geopark through linkages with existing Jeju Island promotion bodies, local communities, travel agents and so on. Linkages between the existing (and future) geosites will be enhanced and formalized. Information and educational materials will be promulgated using printed and multi-media materials as well as advertising campaigns, websites and so on. The Jeju Island Geopark homepage can be found at http://geopark.jeju.go.kr.

10.6.3 Promoting Education

This important Action Plan will build upon the ideas discussed. It will be developed in association with education aut-

horities, specialists and tourism operators. It will emphasize geopark precepts and philosophies and will include concepts for enhanced guiding and interpretation and for the development of geologically based tours. Many Jeju Island elementary and junior high schools already have active curricula which include environmental education as well as promoting of the geological and other heritage of Jeju Island. The Jeju Island Geopark will assist in further developing these programs.

10.6.4 Developing Partnerships

An important function of geopark concept is to create and develop partnerships between geosites, management authorities, scientists, education authorities, local communities and businesses. As an example, the provincial government has recently established a Memorandum of Understanding with the Korean Geological Society to promote and develop the Jeju Island Geopark and to support its inclusion in the Global Geoparks Network. The government of the Jeju Special Self-Governing Province already has many partnerships across the community and this Action Plan will develop protocols and standards for the Jeju Island Geopark 'brand' to ensure that the best and most productive partnerships are achieved. Such standards will include agreements to participate in environmental programs such as recycling, sustainable practices, geopark training activities and so on.

As examples of these standards partnerships will be developed with the following groups:

(1) schools and universities would be encouraged and expected to actively include global geopark principles and practices in their geoscience curricula; (2) hotels and restaurants would actively promote the Jeju Island Geopark and the geosites as well as adopting 'green' environmentally friendly practices; (3) tourist attractions would cross-promote the Jeju Island Geopark and the geosites with the Jeju Special Self-Governing agencies with communications media of appropriate quality; and (4) non-government organizations and local communities would be encouraged to adopt and promote recycling and anti-littering targets and to actively assist with the management and protection of the geosites.

10.6.5 Involvement of the Community and Non-Government Organizations

Various community groups and NGOs are already supporting the Jeju Island Geopark through provision of volunteer georangers and networks such as the Olle walking track system. This Action Plan will seek to expand and further develop links with community bodies. The Department of Geoparks will develop training and information programs for these groups.

10.6.6 Code of Ethics for Visitors and Researchers

This Action Plan will develop codes of ethical practice for visitors, geosite workers and researchers. It will cover such areas as access restrictions, littering, collecting, research protocols and methods, and permit systems for research. They will be promulgated via brochures, information panels and websites.

10.6.7 Training of Managers and Guides

There will be an increased need to have more knowledgeable and better trained employees and volunteer personnel. An Action Plan to enhance training processes and opportunities will be developed and implemented as a high priority.

10.6.8 Promoting Research

Whilst there is already a very considerable body of research on Jeju Island and its geology and geomorphology there are many avenues for ongoing research. There is also the need to investigate further geosite proposals within the Jeju Island Geopark. Promotion of research is an important function of geoparks and this Action Plan will enhance linkages with

academic institutions and specialists to optimize research outcomes and, most importantly, ensure that the results of research are conveyed in an appropriate manner so that they can be used in geosite education and interpretation.

10.7 Monitoring Indicators and Monitoring Plan

Development of clearly identifiable and useful, monitoring indicators for geosite condition, infrastructure condition, visitor satisfaction and of the success or otherwise of knowledge conveyance is a clear requirement. This multi-facetted Action Plan will address these issues in two sub-plans physical matters on one hand (Fig. 10.2) and social conditions on the other. Together they will make up procedures for the ongoing reporting on the Jeju Island Geopark objectives and achievements and effective monitoring plan. It will be necessary to consider the potential impacts of climate change in the monitoring Action Plan. Potential monitoring indicators are set out in the tables below (Table 10.3, 10.4 and 10.5). Effective ongoing maintenance requires the development of protocols, schedules and checklists to ensure the geosites maintain or improve their current condition. This Action Plan will ensure that these are defined and utilized in the better management of the Jeju Island Geopark.

Table 10.3 Monitoring indicators for Majang Cave Geosite (show cave area)

Indicators	Periodicity	Report	Investigator
Daily monitoring list by local management office			
Garbage collection	Daily	Monthly	Management office of the Manjang Cave, JSSGP (Jeju Special Self-Governing Province)
Visitors	Daily	Monthly	
Number of visitors			
Number of international visitors (nationalities)			
Number of students by group tour			
Visitor satisfaction			
Pollution	Daily	Annual	Management office of the Manjang Cave, JSSGP
Lampenflora			
Dust			
Lighting	Daily	Annual	Management office of the Manjang Cave, JSSGP
Direction of lighting into eyes			
Cleaning shields and lamps			
Monitoring by specialists			
Manual air monitoring	Monthly	Annual	JSSGP advised by cave research institute
Temperature			
Humidity			
Carbon dioxide contents			
Radon contents (4 times/year, seasonally)			
Water monitoring	Annual	Annual	
Water quality			
Water level (after rain)			
Automatic monitoring of temperature, humidity and carbon dioxide contents	Monthly	Annual	JSSGP advised by cave research institute
Photo monitoring	Annual	Annual	JSSGP advised by cave research institute
Lava speleothems			
Microtopograhic features			
Safety controls	Annual	Annual	JSSGP advised by cave research institute
Electrical safety			
Infrastructure safety			
Stability of roof and rockfall			
Crack widening check	Annual	Annual	
Cave fauna monitoring	twice/ year (winter, summer)	Annual	JSSGP advised by cave research institute
Electrical safety check investigation	Every 5 years	Every 5 years	JSSGP advised by specialsits
Safety check investigation of infrastructure	Every 5 years	Every 5 years	JSSGP advised by specialsits
3D scan of the show cave area	Every 10–20 years	Every 10–20 years	JSSGP advised by specialsits
Wild cave area			
Cave water quality	Annual	Annual	JSSGP advised by specialISts
Air monitoring	Seasonal	Annual	

10.7 Monitoring Indicators and Monitoring Plan

Table 10.3 (Continued)

Indicators	Periodicity	Report	Investigator
(T, H, CO$_2$, Radon contents)			
Ceiling stability and rockfall	Seasonal	Annual	
Crack checking	Seasonal	Annual	
Photomonitoring	Every 5 years	Every 5 years	
Three entrances			
Significant lava speleothems			
3D scan of the wild cave area	Every 10–20 years	Every 10–20 years	
Soil and sediments transported into the cave	Every 5 years	Every 5 years	
Cave fauna	Every 5 years	Every 5 years	
Outside			
Air monitoring	Daily	Monthly	JSSGP advised by specialsits
Precipitation & snow fall	If needed	Annual	
Photomonitoring	Annual	Annual	
near three entrances			
Significant lava landforms			
Seismicity	If needed	Annual	
Vibration by automobiles	Every 2 years	Every 2 years	
Vegetation	Every 5 years	Every 5 years	

Table 10.4 Monitoring Indicators for Mt. Halla, Cheonjiyeon Waterfall and Sanbangsan Geosites

Indicators	Periodicity	Report	Geosite	Monitoring investigator
Daily monitoring list by local management office				
Garbage collection	Daily	Monthly	All	Local management office
Visitors	Daily	Monthly		
Number of visitors				
Number of international visitors				
Number of students by group tour				
Visitor satisfaction				
Safety of infrastructure	Monthly	Annual	All	
Damage of trails	Monthly	Annual	Mt. Halla	
Rockfall near Sanbanggulsa	Daily	Monthly	Sanbangsan	
Rockfall before the waterfall	Monthly	Annual	Cheonjiyeon	
Landform change above the waterfall	Monthly	Annual	Cheonjiyeon	
Water level	Daily	Annual	Cheonjiyeon	
Rocks transported by the stream	Daily	Annual	Cheonjiyeon	
Water quality (color, smell)	Daily	Annual	Cheonjiyeon	
Monitoring by Mt. Halla Research Institute				
Air monitoring (T, H & Wind velocity)	Daily	Monthly	All	Mt. Halla Research Institute
Precipitation	Daily	Monthly		
Snow depth	If needed	Annual		

Table 10.4 (Continued)

Indicators	Periodicity	Report	Geosite	Monitoring investigator
Water quality check in Baeknokdam	Annual	Annual	Mt. Halla	
Discharge & water quality in Hallasan	Annual	Annual		
Soil quality	Annual	Annual	All	
Photomonitoring	Annual	Annual	All	
Manmade landform change				
Natural landform change				
Valley erosion rate				
Photomonitoring of columnar joints	Annual	Annual	Mt. Halla, Sanbangsan	
Monitoring by specialists				
Seismicity	If needed	Annual	All	JSSGP advised by specialists
Volcanic activity	If needed	Annual		
Distribution of animals	Every 3–5 years	Every 3–5 years	All	
Distribution of plants	Every 3–5 years	Every 3–5 years		
Safety of infrastructure	Every 5 years	Every 5 years	All	
Baeknokdam	Every 5 years	Every 5 years	Mt. Halla	
Slope failure				
Soil erosion & sedimentation				
Monitoring morphological features	Every 5 years	Every 5 years	Mt. Halla	
Collapse of trachytic domes				
Collapse of columnar joints				
Cliff retreat	Every 5 years	Every 5 years	Cheonjiyeon	
Valley erosion rate	Annual	Annual	Mt. Halla	
Calculation of maximum capacity of tourists	Annual	Annual	All	

Table 10.5 Monitoring Indicators for Seongsan Ilchunbong Tuff Cone (Seongsan), Seoguipo Formation, Suweolbong Tuff Ring (Suweolbong), Yongmeori Tuff Ring (Yongmeori) and Daepodong Columnar Joints (Daepodong) Geosites

Indicators	Periodicity	Report	Geosite	Monitoring investigator
Daily monitoring list by local management office				
Garbage collection	Daily	Monthly	All	Local management office
Visitors	Daily	Monthly		
Number of visitors				
Number of international visitors				
Number of students by group tour				
Visitor satisfaction				
Safety of infrastructure	Monthly	Annual	All	
Damage of trails	Monthly	Annual	Seoguipo, Suweolbong, and Yongmeori	

10.7 Monitoring Indicators and Monitoring Plan

Table 10.5 (Continued)

Air monitoring (T, H & Wind velocity)	Daily	Monthly	All	
Precipitation	Daily	Monthly	All	
Snow depth	If needed	Annual	All	
Photomonitoring	Annual	Annual	All	
Manmade landform change				
Natural landform change				
Collapse of vertical cliffs				
Monitoring by specialists				
Soil & sediment accumulation	Annual	Annual	Seongsan	JSSGP advised by specialists
Soil quality	Annual	Annual	All	
Photomonitoring of columnar joints	Annual	Annual	Daepodong	
Seismicity	If needed	Annual	All	
Monitoring of plants (invasion, endangered species)	Annual	Annual	Seongsan	
Soil erosion or accumulation near the crater	Annual	Every 5 years	Seongsan	
Safety of infrastructure	Annual	Every 5 years	All	
Fossil conservation	Annual	Annual	Seoguipo	
Erosion by wave actions and typhoons	Annual	Annual	All	
Cliff retreat	Every 5 years	Every 5 years	Cheonjiyeon	
Vegetation cover on outcrop	Annual	Annual	All	
Calculation of maximum capacity of tourists	Annual	Annual	All	

Fig. 10.1 Monitoring of the Seongsan Ilchulbong Tuff Cone by LiDar. This monitors the changes of vegetation and erosion with a resolution of 50 cm in height and 15 cm in width

Fig. 10.2 Photomonitoring of the outcrop in the Seongsan Ilchulbong Tuff Cone geosite. Photos were taken in 1992 and 2009

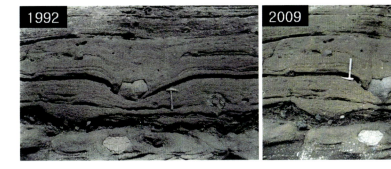

Fig. 10.3 Monitoring of rockfall in the lava tube cave near the road

References

Brenna M, Cronin SJ, Smith IEM, Sohn YK, Németh K (2010) Mechanisms driving polymagmatic activity at a monogenetic volcano, Udo, Jeju Island, South Korea. Contrib Mineral Petr 160:931–950

Cheong CS, Choi JH, Sohn YK, Kim JC, Jeong GY (2007) Optical dating of hydromagmatic volcanoes on the southwestern coast of Jeju Island, Korea. Quat Geochronol 2:266–271

Chough SK, Sohn YK (1990) Depositional mechanics and sequences of base surges, Songaksan tuff ring, Cheju Island, Korea. Sedimentology 37:1115–1135

Haraguchi K (1931) Geology of Cheju Island. Bull Geol Surv Korea 10:1–34. (in Japanese with English abstract)

Hwang SK, Ahn US, Lee MW, Yun SH (2005) Formation and internal structures of the Geomunoreum lava tube system in the northeastern Jeju Island. J Geol Soc Korea 41:385–400

Kang S (2003) Benthic foraminiferal biostratigraphy and paleoenvironments of the Seogwipo Formation, Jeju Island, Korea. J Paleontol Soc Korea 19:63–153

Kang S, Jung KK, Yoon S (1999) Benthic foraminiferal fauna from the Seoguipo Formation of Cheju Island, Korea. J Paleontol Soc Korea 15:95–108

Kim BK (1972) A stratigraphic and paleontologic study of the Seogwipo Formation. The Memoir for Prof. Chi Moo Son's Sixtieth Birthday, pp 167–187

Kim KH, Tanaka T, Suzuki K, Nagao K, Park EJ (2002) Evidences of the presence of old continental basement in Cheju volcanic Island, South Korea, revealed by radiometric ages and Nd-Sr isotopes of granitic rocks. Geochem J 36:421–441

Koh GW (1997) Characteristics of the groundwater and hydrogeologic implications of the Seoguipo Formation in Cheju Island. Ph.D. Thesis, Pusan National University, Pusan, p 326

Koh GW, Park JB, Park YS (2008) The study on geology and volcanism in Jeju Island (I): petrochemistry and 40Ar/39Ar absolute ages of the subsurface volcanic rock cores from boreholes in the eastern lowland of Jeju Island. Econ Environ Geol 41:93–113 (in Korean with English abstract)

Koh J-S, Yun S-H, Hong H-C (2005) Morphology and petrology of Jisagae columnar joint on the Daepodong basalt in Jeju Island, Korea. J Petrol Soc Korea 14:212–225

Koh J-S, Yun S-H, Hyeon G-B, Lee M-W, Gil Y-W (2005) Petrology of the basalt in the Udo monogenetic volcano, Jeju Island. J Petrol Soc Korea 14:45–60

Lee DY, Yun SK, Kim JY, Kim YJ (1988) Quaternary geology of the Jeju Island. KR-87-29, Korea Institute of Energy and Resources, Taejon, pp 233–278

Lee MW, Won CK, Lee DY, Park GH (1994) Stratigraphy and petrology of volcanic rocks in southern Cheju Island, Korea. J Geol Soc Korea 30:521–541

Li B, Park BK, Kim DS, Woo HJ (1999) The geologic age and paleoenvironment of the lower Seoguipo Formation, Cheju Island, Korea. Geosci J 3:181–190

Park KH, Lee BJ, Kim JC, Cho DL, Lee SR, Choi HI, Park DW, Lee SR, Choi YS, Yang DY, Kim JY, Seo JY, Sin HM (2000) Explanatory note of the Jeju (Baekado, Jinnampo) sheet (1:250,000). Korea Institute of Geoscience and Mineral Resources, Taejon

Reading HG (ed) (1996) Sedimentary environments: processes, facies and stratigraphy. Blackwell Science, Oxford, p 688

Sigurdsson H, Houghton BF, McNutt SR, Rymer H, Stix J (eds) (2000) Encyclopedia of volcanoes. Academic, San Diego, p 1417

Sohn YK (1996) Hydrovolcanic processes forming basaltic tuff rings and cones on Cheju Island, Korea. Geol Soc Am Bull 108:1199–1211

Sohn YK, Chough SK (1989) Depositional processes of the Suwolbong tuff ring, Cheju Island (Korea). Sedimentology 36:837–855

Sohn YK, Chough SK (1992) The Ilchulbong tuff cone, Cheju Island, South Korea: depositional processes and evolution of an emergent, Surtseyan-type tuff cone. Sedimentology 39:523–544

Sohn YK, Chough SK (1993) The Udo tuff cone, Cheju Island, South Korea: transformation of pyroclastic fall into debris fall and grain flow on a steep volcanic cone slope. Sedimentology 40:769–786

Sohn YK, Park KH (2004) Early-stage volcanism and sedimentation of Jeju Island revealed by the Sagye borehole, SW Jeju Island, Korea. Geosci J 8:73–84

Sohn YK, Park KH (2005) Composite tuff ring/cone complexes in Jeju Island, Korea: possible consequences of substrate collapse and vent migration. J Volcanol Geotherm Res 141:157–175

Sohn YK, Park KH (2007) Phreatomagmatic volcanoes of Jeju Island, Korea: IAVCEI-CEV-CVS Field Workshop, Jeju Island, Korea. OB Communications, Daejeon, p 83

Sohn YK, Yoon S-H (2010) Shallow-marine records of pyroclastic surges and fallouts over water in Jeju Island, Korea, and their stratigraphic implications. Geology 38:763–766

Sohn YK (1992) Depositional models of basaltic tuff rings and tuff cones in Cheju Island, Korea. Ph.D. Thesis, Seoul National University, Seoul, p 210

Sohn YK (1995) Structures and sequences of the Yongmeori tuff ring, Cheju Island, Korea: Sequential deposition from shifting vents. J Geol Soc Korea 31:57–71

Sohn YK, Park JB, Khim BK, Park KH, Koh GW (2002) Stratigraphy, petrochemistry and Quaternary depositional record of the Songaksan tuff ring, Jeju Island, Korea. J Volcanol Geotherm Res 19:1–20

Sohn YK, Park JB, Yoon SH (2008) Primary versus secondary and subaerial versus submarine hydrovolcanic deposits in the subsurface of Jeju Island, Korea. Sedimentology 55:899–924

Tamanyu S (1990) The K-Ar ages and their stratigraphic interpretation of the Cheju Island volcanics, Korea. Bull Geol Serv Japan 41:527–537

Vespermann D, Schmincke H-U (2000) Scoria cones and tuff rings. In: Sigurdsson H, Houghton BF, McNutt SR, Rymer H, Stix J (eds) Encyclopedia of Volcanoes. Academic, San Diego, pp 683–694

Walker RG, James NP (eds) (1992) Facies Models: Response to Sea Level Change. Geological Society of Canada, p 409

White JDL, Houghton B (2000) Surtseyan and related phreatomagmatic eruptions. In: Sigurdsson H, Houghton BF, McNutt SR, Rymer H, Stix J (eds) Encyclopedia of Volcanoes. Academic, San Diego, pp 495–511

Wohletz K, Sheridan M (1983) Hydrovolcanic eruptions II. Evolution of basaltic tuff rings and tuff cones. Am J Sci 283:385–413

Won JK, Matsuda J, Nagao K, Kim KH, Lee MW (1986) Paleomagnetism and radiometric age of trachytes in Jeju Island, Korea. J Korean Inst Min Geol 19:25–33

Yi S, Yun H, Yoon S (1998) Calcareous nannoplankton from the Seoguipo Formation of Cheju Island, Korea and its paleoceanographic implications. Paleontol Res 2:253–265

Yokoyama M (1923) On some fossil shells from the island of Saishu in the Strait of Tsushima. J Coll Sci Imp Univ Tokyo 44:1–9 (in Japanese with English abstract)

Yoon S (1988) The Seoguipo molluscan fauna of Jeju Island, Korea. Saito Ho-on Kai Special Publication, Japan, pp 539–545

Printed by Publishers' Graphics LLC
MLSI130719.15.13.137